Carl Zimmer

The Free Press

New York   London   Toronto   Sydney   Singapore

PARASITE REX

Inside the Bizarre World of Nature's Most Dangerous Creatures

THE FREE PRESS
A Division of Simon & Schuster, Inc.
1230 Avenue of the Americas
New York, NY 10020

Copyright © 2000 by Carl Zimmer

Designed by Deirdre C. Amthor

Manufactured in the United States of America

1   3   5   7   9   10   8   6   4   2

Library of Congress Cataloging-in-Publication Data

Zimmer, Carl
Parasite rex : inside the bizarre world of nature's most dangerous creatures /
Carl Zimmer.
p.   cm.
Includes bibliographical references and index.
1. Parasites. I. Title

QL757 .Z56 2000
591.7'857—dc21
00-037593

ISBN 0-684-85638-7

Thanks to Andreas Schmidt-Rhaesa for the photograph of the nematomorph (photo
insert, p.10) originally published in Schmidt-Rhaesa, et al., *Journal of Natural History*
34 (2000): p. 338, by Taylor & Francis, Ltd.

Also by Carl Zimmer

*At the Water's Edge: Macroevolution
and the Transformation of Life*

Contents

Contents

Prologue:
A Vein Is a River

The boy in the bed in front of me was named Justin, and he didn't want to wake up. His bed, a spongy mat on a metal frame, sat in a hospital ward, a small concrete building with empty window frames. The hospital was made up of a few of these buildings, some with thatched roofs, in a wide dusty courtyard. It felt more like a village than a hospital to me. I associate hospitals with cold linoleum, not with goat kids in the courtyard, punching udders and whisking their tails, not with mothers and sisters of patients tending iron pots propped up on little fires under mango trees. The hospital was on the edge of a desolate town called Tambura, and the town was in southern Sudan, near the border with the Central African Republic. If you were to travel out in any direction from the hospital, you would head through little farms of millet and cassava, along winding paths through broken forests and swamps, past concrete-and-brick funeral domes topped with crosses, past termite mounds shaped like

Prologue

giant mushrooms, past mountains covered in venomous snakes, elephants, and leopards. But since you're not from southern Sudan, you probably wouldn't have traveled out in any direction, at least not when I was there. For twenty years a civil war had been lingering in Sudan between the southern tribes and the northerners. When I visited, the rebels had been in control of Tambura for four years, and they decreed that any outsiders who arrived on the weekly prop plane that landed on its muddy airstrip could travel only with rebel minders, and only in the daytime.

Justin, the boy in the bed, was twelve years old, with thin shoulders and a belly that curved inward like a bowl. He wore khaki shorts and a blue-beaded necklace; on the window ledge above him was a sack woven from reeds and a pair of sandals, each with a metal flower on its thong. His neck was so swollen that it was hard to tell where the back of his head began. His eyes bulged in a froglike way, and his nostrils were clogged shut.

"Hello, Justin! Justin, hello?" a woman said to him. There were seven of us there at the boy's bedside. There was the woman, an American doctor named Mickey Richer. There was an American nurse named John Carcello, a tall middle-aged man. And there were four Sudanese health workers. Justin tried to ignore all of us, as if we'd all just go away and he could go back to sleep. "Do you know where you are?" Richer asked him. One of the Sudanese nurses translated into Zande. He nodded and said, "Tambura."

Richer gently propped him up against her side. His neck and back were so stiff that when she lifted him he rose like a plank. She couldn't bend his neck, and as she tried, Justin, his eyes barely open, whimpered for her to stop. "If this happens," she said emphatically to the Sudanese, "call a doctor." She was trying to hide her irritation that they hadn't called her already. The boy's stiff neck meant that he was at the edge of death. For weeks his body had been overrun with a single-celled parasite, and the medicine Richer was giving him wasn't working. And there were a hundred other patients in Richer's hospital, all of whom had the same fatal disease, called sleeping sickness.

I had come here to Tambura for its parasites, the way some people go to Tanzania for its lions or Komodo for its dragons. In New York, where I live, the word *parasite* doesn't mean much, or at least not much in particular. When I'd tell people there I was studying parasites, some would say, "You mean tapeworms?" and some would say, "You mean ex-wives?" The word is slippery. Even in scientific circles, its definition can slide around. It can mean anything that lives on or in another organism at the expense of that organism. That definition can include a cold virus or the bacteria that cause meningitis. But if you tell a friend with a cough that he's harboring parasites, he may think you mean that there's an alien sitting in his chest, waiting to burst out and devour everything in sight. Parasites belong in nightmares, not in doctors' offices. And scientists themselves, for peculiar reasons of history, tend to use the word for everything that lives parasitically *except* bacteria and viruses.

Even in that constrained definition, parasites are a vast menagerie. Justin, for example, was lying in his hospital bed on the verge of death because his body had become home to a parasite called a trypanosome. Trypanosomes are single-celled creatures, but they are far more closely related to us humans than to bacteria. They got into Justin's body when he was bitten by a tsetse fly. As the tsetse fly drank his blood the trypansomes poured in. They began to steal oxygen and glucose from Justin's blood, multiplied and eluded his immune system, invaded his organs, and even slipped into his brain. Sleeping sickness gets its name from the way trypansomes disrupt people's brains, wrecking their biological clock and turning day to night. If Justin's mother hadn't brought him to the Tambura hospital, he would certainly have died in a matter of months. Sleeping sickness is a disease without pardon.

When Mickey Richer had come to Tambura four years ear-

Prologue

lier, there were hardly any cases of sleeping sickness, and people generally thought of it as a disease that was fading into history. That wasn't always the case. For thousands of years, sleeping sickness has threatened people in the range of the tsetse fly: a wide swath of Africa south of the Sahara. A version of the disease also attacked cattle and kept vast regions of the continent free of domesticated animals. Even now, over 4.5 million square miles are off limits to cattle in Africa because of sleeping sickness, and even where people do raise cattle, 3 million die of sleeping sickness each year. When Europeans colonized Africa, they helped trigger giant epidemics by forcing people to stay and work in tsetse-infested places. In 1906, Winston Churchill, who was the colonial undersecretary at the time, told the House of Commons that one sleeping sickness epidemic had reduced the population of Uganda from 6.5 million to 2.5 million.

By World War II, scientists had discovered that drugs effective against syphilis could also eradicate trypanosomes from the body. They were crude poisons, but they worked well enough to make the parasites sink back down to low levels if doctors carefully screened places thick with tsetse flies and treated the sick. There would always be sleeping sickness, but it would be an exception, not the rule. Campaigns against sleeping sickness during the 1950s and 1960s were so effective that scientists talked of eliminating the disease in a matter of years.

But war, crumbling economies, and corrupt governments let sleeping sickness come back. In Sudan the civil war drove away Belgian and British doctors from Tambura County; they had been keeping a careful watch for outbreaks. Not far from Tambura, I visited an abandoned hospital that had had its own sleeping sickness ward; now it is filled with wasps and lizards. As the years passed, Richer watched her load of sleeping sickness cases rise, first to 19, then to 87, then to hundreds. She ran a survey in 1997 and estimated from it that 20 percent of the people in Tambura County—12,000 Sudanese—carried sleeping sickness.

Prologue

That year Richer launched a counteroffensive, hoping to fight back the parasite at least in Tambura county. For people who were still in the early stages of the disease, ten days of injections in the buttocks with the drug pentamidine was enough. For those like Justin who had the parasites in their brains, a harsher course was necessary. They needed stronger stuff that could kill the parasite outright in their brain—a brutal potion known as melarsoprol. Melarsoprol is made of 20 percent arsenic. It can melt ordinary plastic IV tubes, so Richer had to have tubes flown in that were as tough as Teflon. If melarsoprol seeps out of a vein, it can turn the surrounding flesh into a swollen, painful mass; then, at the very least the drugs have to be stopped for a few days, and at worst the arm may have to be amputated.

When Justin arrived at the hospital, he already had parasites in his brain. The nurses gave him injections of melarsoprol for three days, and the medicine wiped out a fair number of the trypanosomes in his brain and spine. But as a result, his brain and spine had been flooded with scraps of dead parasite tissue, driving his immune cells from a torpor to a frenzy. They shot out blasts of poisons, which scorched Justin's brain. The inflammation they triggered was squeezing it like a vise.

Now Richer prescribed steroids for Justin to try to bring the swelling down. Justin whimpered remotely as the needleful of steroids went into his arm, his eyes closed as if he were deep in a bad dream. If he was lucky, the steroids would take pressure off his brain. The next day would tell: either he would be better or he would be dead.

Before I arrived at Justin's bedside, I had been traveling with Richer for a few days, watching her at work. We had gone to villages where her staff was spinning blood in centrifuges, looking for the signature of the parasite. We had driven for hours to get to another clinic of hers, where people were getting spinal taps to see if the trypanosomes were on their way to the brain. We had made the rounds of the Tambura hospital,

Prologue

seeing other patients: little children who had to be held down for injections as they screamed, old women bearing up silently as the medicine burned into their veins, a man made so crazy by the medicine that he had taken to attacking people and needed to be tied to a post. And from time to time—and now, as I looked at Justin—I tried to see the parasites inside them. It brought to mind that old movie *Fantastic Voyage*, in which Raquel Welch and her fellow crewmates climb into a submarine that is then squinched down to microscopic size. They are injected into a vein in a diplomat's body so that they can travel through his circulatory system to his brain and save him from a life-threatening wound. I had to enter that world, made of underground rivers, where the currents of blood follow ever-smaller branches of arteries until they pass back around into the veins, joining up to larger veins until they reach the surging heart. Red blood cells bounced and rolled along, squeezing through capillaries and then rebounding to their original puck shapes. White blood cells used their lobes to crawl into the vessels through lymphatic ducts, like doorways disguised as bookshelves in a house. And among them traveled the trypanosomes. I have looked at trypanosomes under a microscope in a Nairobi laboratory, and they are quite beautiful. Their name comes from *trypanon*, the Greek word for an augur. They are about twice as long as a red blood cell, silvery under a microscope. Their bodies are flat, like a strip, but as they swim they spin like drill bits.

Parasitologists who spend enough time looking at trypanosomes in laboratories tend to fall in love with them. In an otherwise sober scientific paper, I came across this sentence: "*Trypanosoma brucei* has many enchanting features that have made this parasite the darling of experimental biologists." Parasitologists watch the trypanosomes as carefully as an ornithologist watches ospreys, while the parasites gulp glucose, while they evade the pursuit of immune cells by tossing off their coat and putting on a new one, while they transform themselves into new forms that can survive in the gut of a fly and then transform back into a form perfectly adapted for human hosts.

Trypanosomes are only one of many parasites inside the people of southern Sudan. If you could travel *Fantastic Voyage*–style through their skin, you would probably come across marble-sized nodules where you'd float past coiled worms as long as snakes and as thin as threads. Called *Onchocerca volvulus*, these animals, male and female, spend their ten-year-long lives in these nodules, making thousands of babies. The babies leave them and travel within the skin, in the hope that they'll get taken up in the bite of a black fly. In the black fly's gut they can mature to their next stage, and the insect can then inject them in the skin of a new host, where they will form a nodule of their own. As the babies swim through a victim's skin they can trigger a violent attack from the immune system. Rather than kill the parasite, though, the immune system puts a rash of leopard spots on the skin of its host. The rash can get so itchy that people may scratch themselves to death. When the worms wander through the outer layer of the eyes, the immune system's scarring can leave a person blind. Since their larvae are aquatic, black flies tend to stay around water, and the disease has thus earned the name *river blindness*. There are some places in Africa where river blindness has claimed the eyes of just about every person over forty.

Then there are Tambura's guinea worms: two-foot-long creatures that escape their hosts by punching a blister through the leg and crawling out over the course of a few days. Then there are filarial worms that cause elephantiasis, which can make a scrotum swell up until it can fill a wheelbarrow. Then there are tapeworms: eyeless, mouthless creatures that live in the intestines, stretching as long as sixty feet, made up of thousands of segments, each with its own male and female sex organs. There are leaf-shaped flukes in the liver and the blood. There are single-celled parasites that cause malaria, invading blood cells and exploding them with a fresh new generation hungry for cells of their own. Stay long enough in Tambura, and people around you turn transparent and become glittering constellations of parasites.

Tambura is not as freakish as it might seem. It's just a place where you can find parasites thriving in humans with particular ease. Most people on Earth carry parasites, even if you set aside bacteria and viruses. Over 1.4 billion people carry the snakelike roundworm *Ascaris lumbricoides* in their intestines; almost 1.3 billion carry blood-sucking hookworms; 1 billion have whipworm. Two or three million die of malaria a year. And many of these parasites are on the rise, not the wane. Richer may be slowing down the spread of sleeping sickness in her little patch of Sudan, but around her it seems to be spreading. It may kill three hundred thousand people a year; it probably kills more people in the Democratic Republic of Congo than AIDS. Parasitically speaking, New York is actually more freakish than Tambura. And if you step back and survey our evolution from an apelike ancestor 5 million years ago, the past century of parasite-free living that some humans have enjoyed is a fleeting reprieve.

I checked in on Justin the following day. He was propped up on his side, eating broth from a bowl. His back was lazily curved along the bed as he ate; his eyes were no longer swollen; his neck was supple again; his nose was clear. He was still exhausted and was far more interested in eating than in talking to strangers. But it was good to see that the fleeting reprieve included him as well.

• • •

Visiting places like Tambura, I began to think of the human body as a barely explored island of life, home to creatures unlike anything in the outside world. But when I remembered that we are just one species out of millions on this planet, the island swelled up to a continent, a planet.

A few months after my trip to Sudan, on a night that wa-

Prologue

vered between muggy and rainy, I walked through a Costa Rican jungle. I held a butterfly net in my hand, and the pockets of my raincoat spilled over with plastic bags. The headlamp on my brow cast a slanted oval on the path in front of me, which a spider crossed twenty feet ahead. Its eight eyes glinted together like a single diamond chip. A giant solitary wasp crawled slowly into its burrow on the side of the path to hide from my glare. The only light beyond my lamp came from distant lightning and the fireflies that glowed for long slow flashes in the trees overhead. The grass gave off the rank odor of jaguar urine.

I walked with seven biologists, led by one scientist named Daniel Brooks. He was about as far from my picture of the intrepid jungle biologist as he could get: heavy frame, a drooping mustache, and big aviator glasses, dressed in a red-and-black jogging suit and sneakers. But as the rest of us passed the time on the walk by talking about how to photograph birds or how to tell the difference between a poisonous coral snake and a harmless mimic, Brooks kept ahead, listening to the peeps and croaks that surrounded us. He stopped suddenly at the side of the path, waving his right hand back and low to shut us up. He moved toward a broad ditch filling with the night's rain and lifted his net slowly. He put one sneaker into the water and then suddenly brought the net down on the far bank. Its pointed end started dancing and punching, and he grabbed the net midway before raising it. With his other hand he took a plastic bag from me and blew it full of air. He transferred a big beige-striped leopard frog into the bag, where it jumped frantically. He knotted the open end of the bag, still fat with air, and wedged the knot under the drawstring of his sweatpants. He started walking down the path again with his bulging frog bag, a transparent sack of gold.

Frogs and toads were everywhere that night. Brooks caught a second leopard frog not far down the path. Tungara frogs drifted in the water, in powerful choruses. Marine toads, some as big as cats, waited until we were close by before taking a single big lazy hop to keep their distance. We walked past blobs of foam as firm as bubble bath, out of which hundreds of tadpoles squirmed into the nearby water. We caught blunt-faced microhylid frogs, with tiny stupid eyes crowded up just over their nostrils and fat low bodies shaped like dollops of chocolate pudding.

For some zoologists, the hunt for their animals would be over at this point. But Brooks wasn't sure yet what he had actually found. He brought the frogs back to the headquarters of the Area de Conservacion de Guanacaste. He left the frogs in their bags overnight, with some water to keep them damp and alive. In the morning, after a breakfast of rice and beans and pineapple juice, he and I went to his lab. The lab consisted of a shed with chicken-wire walls on two sides.

"The assistants here call it the *jaula*," said Brooks. There was a table in the middle of the shed that held dissecting microscopes, and woolly bears and beetles crawled across its concrete floor. A mud wasp nest hung from the light cord. Outside, beyond the vines that surrounded the shed, a howler monkey roared in the trees. *Jaula* means "jail" in Spanish. "They say that we have to stay in here or we'd kill all their animals."

Brooks took out a leopard frog from the bag and dispatched it with a sharp thwack on the edge of the sink. It was dead in an instant. He laid it on the table and began snipping its belly open. He used tweezers to pull the guts delicately free of the frog's trunk. He put the organs into a broad petri dish and put the husk of the frog under a microsope. During the previous three summers, Brooks had looked into the insides of eighty species of reptiles, birds, and fish at Guanacaste. He had started making a list of every parasite species that lives in the reserve. There are so many different kinds of parasites within the animals and plants of the world that no one had ever dared such a thing in a place the size of Guanacaste. He adjusted the lights on their long black stalks, two curious snakes looking at the dead frog. "Ah," he said, "here we go."

He had me look: a filarial worm—a relative of guinea

worms in humans—had come wandering out of its home in one of the veins in the frog's back. "It's probably transmitted by mosquitoes that feed on the frogs," Brooks explained. He pulled it out intact and dropped it in a dish of water. By the time he had gotten a dish of acetic acid (industrial-strength vinegar) to fix it in, the parasite had exploded into a white froth. But Brooks was able to get another one out untorn and into the acid unexploded, where it straightened out, ready to be preserved for decades.

That was the first of many parasites we looked at. A string of flukes came out, like a writhing necklace, from another vein. The kidneys carried another species that only mature when the frog is eaten by a predator like a heron or a coati. The lungs of this frog were clear, although often the frogs here will have parasites in their lungs as well. They get several malarias in their blood, even get flukes in their esophagus and ears. "Frogs are parasite hotels," Brooks said. He worked apart the intestines, slitting them carefully so that he wouldn't snip any parasites inside. He found another species of fluke, a tiny fleck that swam across the microscope's view. "If you didn't know what to look for, you'd think it was garbage. It goes from a snail to a fly, which is then eaten by a frog." The fluke has to share this particular set of intestines with a trichostrongylid worm that takes a more direct route to get there, burrowing straight into the frog's gut.

Brooks pushed the dish out from under the microscope. "That was real disappointing, guys," he said. I think he was addressing the parasites. I was pretty overwhelmed by all the creatures I'd just seen in one animal, but Brooks knew that a single frog species may have a dozen species inside it, and he wanted me to see as many as I could. He spoke to the frog: "Let's hope your compadre has more."

He reached into the bag for the second leopard frog. This one had two toes missing from its front left foot. "That means he escaped from a predator that wasn't as successful as me," Brooks said, and dispatched it with another swift thwack. When

Prologue

he got its open belly under the microscope, he said "Oh!" with a sudden brightness. "This is nice. Sorry. Relatively speaking, this is nice." He had me look through the eyepieces. Another fluke, this one called a gorgoderid, for its resemblance to the writhing snakes on Medusa's head, was twisting out of the frog's bladder. "They live in freshwater clams. This tells me this frog has been somewhere where there are clams, which need a guaranteed water supply, sandy bottom, calcium-rich soil. And its second host is a crayfish, so the habitat has to support clams, crayfish, and frogs, and do it year round. Where we caught him yesterday is not where he comes from." He moved on to its intestines. "Here's a nice little vignette"—nematodes alongside flukes that form cysts on the frog's skin. When the frog sheds its skin, it eats it, thereby infecting itself. The flukes were acrobatic sacs of eggs.

Cheered up now, Brooks moved on to a blobby microhyalid frog. "Oh my, you've brought me luck," he says, looking inside it. "This thing must have a thousand pinworms. Holy cow, this guy is crawling." In the pinworm soup there were squirming iridescent protozoa, single-celled giants that were almost as big as the multicellular worms.

A few of the parasites we saw already have names, but most are new to science. For now, Brooks went to his computer and typed in vague descriptors—nematode, tapeworm—that would be honed down by himself or some other parasitologist who would come up with a Latin name. The computer carried in it the records of other parasites Brooks had recorded over the years, including some of the ones I had watched dissected over the course of the previous few days. There were the iguanas with their tapeworms, the turtle with an ocean of pinworms. Just before my arrival, Brooks and his assistants had opened up a deer and found a dozen species living in or on it, including nematodes that live only in the deer's Achilles tendon and flies that lay their eggs in the deer's nose. (Brooks calls these last ones the snot bots.)

Prologue

Even within this one reserve, Brooks was probably not going to be able to count every parasite. Brooks is an expert on the parasites of vertebrates as parasites are traditionally defined—in other words, excluding the bacteria and viruses and fungi. When I visited him, he had identified about three hundred of these parasites, but he estimated there would be eleven thousand in total. Brooks doesn't study the thousands of species of parasitic wasps and flies that live in the forest, devouring insects from within and keeping them alive till the last moment of their feast. He doesn't study the plants that parasitize other plants, stealing the water their hosts pump from the ground and the food they make out of air and sun. He doesn't study fungi, which can invade animals, plants, or even other fungi. He can only hope that other parasitologists will join him. They are spread thin over their subjects. Every living thing has at least one parasite that lives inside it or on it. Many, like leopard frogs and humans, have many more. There's a parrot in Mexico with thirty different species of mites on its feathers alone. And the parasites themselves have parasites, and some of those parasites have parasites of their own. Scientists such as Brooks have no idea just how many species of parasites there are, but they do know one dazzling thing: parasites make up the majority of species on Earth. According to one estimate, parasites may outnumber free-living species four to one. In other words, the study of life is, for the most part, parasitology.

The book in your hand is about this new study of life. Parasites have been neglected for decades, but recently they've caught the attention of many scientists. It has taken a long time for scientists to appreciate the sophisticated adaptations parasites have made to their inner world, because it is so hard to get a glimpse of it. Parasites can castrate their hosts and then take over their minds. An inch-long fluke can fool our complex immune system into thinking it is as harmless as our own blood. A wasp can insert its own genes into the cells of a caterpillar to shut down the caterpillar's immune system. Only now are scientists thinking seriously about how parasites may be as important to ecosystems as lions and leopards. And only now are they realizing that parasites have been a dominant force, perhaps the dominant force, in the evolution of life.

Or perhaps I should say in the minority of life that is not parasitic. It takes a while to get used to that.

1

Nature's Criminals

Nature is not without a parallel strongly suggestive of our social perversions of justice, and the comparison is not without its lessons. The ichneumon fly is parasitic in the living bodies of caterpillars and the larvae of other insects. With cruel cunning and ingenuity surpassed only by man, this depraved and unprincipled insect perforates the struggling caterpillar, and deposits her eggs in the living, writhing body of her victim.

—John Brown, in *Parasitic Wealth or Money Reform: A Manifesto to the People of the United States and to the Workers of the World* (1898)

In the beginning there was fever. There was bloody urine. There were long quivering strings of flesh that spooled out of the skin. There was a sleepy death in the wake of biting flies.

Parasites made themselves, or at least their effects, known thousands of years ago, long before the name parasite—*parasitos*—was created by the Greeks.

The word literally means "beside food," and the Greeks originally had something very different in mind when they used it, referring to officials who served at temple feasts. At some point the word slipped its etymological harness and came to mean a hanger-on, someone who could get the occasional meal from a nobleman by pleasing him with good conversation, delivering messages, or doing some other job. Eventually the parasite became a standard character in Greek comedy, with his own mask. It would be many centuries before the word would cross over to biology, to define life that drains other lives from within. But the Greeks already knew of biological parasites. Aristotle, for instance, recognized creatures that lived on the tongues of pigs, encased in cysts as tough as hailstones.

People knew about parasites elsewhere in the world. The ancient Egpytians and Chinese prescribed different sorts of plants to destroy worms that lived in the gut. The Koran tells its readers to stay away from pigs and from stagnant water, both sources of parasites. For the most part, though, this ancient knowledge has only left a shadow on history. The quivering strings of flesh—now known as guinea worms—may have been the fiery serpents that the Bible describes plaguing the Israelites in the desert. They certainly plagued much of Asia and Africa. They couldn't be yanked out at one go, since they would snap in two and the remnant inside the body would die and cause a fatal infection. The universal cure for guinea worm was to rest for a week, slowly winding the worm turn by turn onto a stick to keep it alive until it had crawled free. Someone figured out this cure, someone forgotten now for perhaps thousands of years. But it may be that that person's invention was remembered in the symbol of medicine, known as the caduceus: two serpents wound around a staff.

As late as the Renaissance, European physicians generally thought that parasites such as guinea worms didn't actually make people sick. Diseases were the result of the body itself lurching out of balance as a result of heat or cold or some other force. Breathing in bad air could bring on a fever called malaria, for example. A disease came with symptoms: it made people cough, put spots on their belly, gave them parasites. Guinea worms were the product of too much acid in the blood, and weren't actually worms at all—they were something made by a diseased body: perhaps corrupted nerves, black bile, elongated veins. It was hard to believe, after all, that something as bizarre as a guinea worm could be a living creature. Even as late as 1824, some skeptics still held out: "The substance in question cannot be a worm," declared the superintending surgeon of Bombay, "because its situation, functions, and properties are those of a lymphatic vessel and hence the idea of its being an animal is an absurdity."

Other parasites were undeniably living creatures. In the intestines of humans and animals, for instance, there were slender snake-shaped worms later named *Ascaris*, and tapeworms—flat, narrow ribbons that could stretch for sixty feet. In the livers of sick sheep were lodged parasites in the shape of leaves, called flukes after their resemblance to flounder (*floc* in Anglo-Saxon). Yet, even if a parasite was truly a living creature, most scientists reasoned, it also had to be a product of the body itself. People carrying tapeworms discovered to their horror that strips of it would pass out with their bowel movements, but no one had ever seen a tapeworm crawl, inch by inch, into a victim's mouth. The cysts that Aristotle had seen in the tongues of pigs had little wormlike creatures coiled up inside, but these were helpless animals that didn't even have sex organs. Parasites, most scientists assumed, must have been spontaneously generated in bodies, just as maggots appeared spontaneously on a corpse, fungus on old hay, insects from within trees.

In 1673, the visible parasites were joined by a zoo of invisible ones. A shopkeeper in the Dutch city of Delft put a few drops of old rainwater under a microscope he had built himself, and he saw crawling globules, some with thick tails, some with paws. His name was Anton van Leeuwenhoek, and although in his day he was never considered anything more than an amateur, he was the first person to lay eyes on bacteria, to see cells. He put everything he could under his microscope. Scraping his teeth, he discovered rod-shaped creatures living on them, which he could kill with a sip of hot coffee. After a disagreeable meal of hot smoked beef or ham, he would put his own loose stool under his lenses. There he could see more creatures—a blob with leglike things that it used to crawl like a wood louse, eel-shaped creatures that would swim like a fish in water. His body, he realized, was a home to microscopic parasites.

Other biologists later found hundreds of different kinds of microscopic creatures living inside other creatures, and for a couple of centuries there was no divide between them and the bigger parasites. The new little worms took many shapes—of frogs, of scorpions, of lizards. "Some shoot forth horns," one biologist wrote in 1699, "others acquire a forked Tail; some assume Bills, like Fowls, others are covered with Hair, or become all over rough; and others again are covered with Scales and resemble Serpents." Meanwhile, other biologists identified hundreds of different visible parasites, flukes, worms, crustaceans, and other creatures living in fish, in birds, in any animal they opened up. Most scientists still held on to the idea that parasites large and small were spontaneously generated by their hosts, that they were only passive expressions of disease. They held on through the eighteenth century, even as some scientists tested the idea of spontaneous generation and found it wanting. These skeptics showed how the maggots that appeared on the corpse of a snake were laid as eggs by flies, and themselves grew into flies.

Even if maggots weren't spontaneously generated, parasites were a different matter. They simply had no way of getting inside a body and so had to be created there. They had never been seen outside a body, animal or human. They could be found in young animals, even in aborted fetuses. Some species could be found in the gut, living happily alongside other organisms that were being destroyed by digestive juices. Others could be found clogging the heart and the liver, without any conceivable way to get into those organs. They had hooks and suckers and other equipment for making their way inside a body, but they would be helpless in the outside world. In other words, parasites were clearly designed to live their entire lives inside other animals, even in particular organs.

Spontaneous generation was the best explanation for parasites, given the evidence at hand. But it was also a profound heresy. The Bible taught that life was created by God in the first week of creation, and every creature was a reflection of His design and His beneficence. Everything that lived today must descend from those primordial creatures, in an unbroken chain of parents and children—nothing could later come squirting into existence thanks to some vital, untamed force. If our own blood could spontaneously generate life, what help did it need from God back in the days of Genesis?

The mysterious nature of parasites created a strange, disturbing catechism of its own. Why did God create parasites? To keep us from being too proud, by reminding us that we were merely dust. How did parasites get into us? They must have been put there by God, since there was no apparent way for them to get in by themselves. Perhaps they were passed down through generations within our bodies to the bodies of our children. Did that mean that Adam, who was created in purest innocence, came into being already loaded with parasites? Maybe the parasites were created inside him after his fall. But wouldn't this be a second creation, an eighth day added on to that first week—"and on the following Monday God created parasites"? Well, then, maybe Adam was created with parasites after all, but in Eden parasites were his helpmates. They ate the food he couldn't fully digest and licked his wounds clean from within. But why should Adam, created not only in innocence but in perfection, need any help at all? Here the catechism seems to have finally fallen apart.

Parasites caused so much confusion because they have life

cycles unlike anything humans were used to seeing. We have the same sorts of bodies as our parents did at our age, as do salmon or muskrats or spiders. Parasites can break that rule. The first scientist to realize this was a Danish zoologist, Johann Steenstrup. In the 1830s he contemplated the mystery of flukes, whose leaf-shaped bodies could be found in almost any animals a parasitologist cared to look at—in the livers of sheep, in the brains of fish, in the guts of birds. Flukes laid eggs, and yet no one in Steenstrup's day had ever found a baby fluke in its host.

They had, however, found other creatures that looked distinctly flukish. Wherever certain species of snails lived, in ditches or ponds or streams, parasitologists came across free-swimming animals that looked like small versions of flukes except that they had great tails attached to their rears. These animals, called cercariae, flicked their tails madly through the water. Steenstrup scooped up some ditch water, complete with snails and cercariae, and kept it in a warm room. He noticed that the cercariae would penetrate the mucus coating the snail's body and shell, drop their tails, and form a hard cyst, which, he said, "arches over them like a small, closely-shut watch glass." When Steenstrup pulled the cercariae out of these shelters, he found that they had become flukes.

Biologists knew that the snails were home to other sorts of parasites as well. There was a creature that looked like a shapeless bag. There was also a little beast they called the King's yellow worm: a pulpy animal that lived in the snail's digestive gland and carried within it what looked like cercariae, all writhing like cats inside a burlap sack. And Steenstrup even found another flukelike creature swimming free, this one not using a missile-shaped tail but instead hundreds of fine hairs that covered its body.

Looking at all these organisms swimming through the water and through the snails—organisms that in many cases had been given their own Latin species names—Steenstrup made an outrageous suggestion. All these animals were different stages and generations of a single animal. The adults laid eggs, which escaped out of their hosts and landed in water, where they hatched into the form covered in fine hairs. The hair-covered form swam through the water and sought out a snail, and once it had penetrated a snail, the parasite transformed itself into the shapeless bag. The shapeless bag began to swell with the embryos of a new generation of flukes. But these new flukes were nothing like the leaf-shaped forms inside a sheep's liver, or even the finely haired form that entered the snail. These were the King's yellow worms. They moved through the snail, feeding and rearing within them yet another generation of flukes—the missile-tailed cercariae. The cercariae emerged from the snail, promptly forming cysts on the snail. From there they somehow got into sheep or another final host, and there they emerged from their cysts as mature flukes.

Here was a way that parasites could appear inside our bodies with no precedent: "An animal bears young which are, and remain, dissimilar to their parent, but bring forth a new generation, whose members either themselves, or in their descendants, return to the original form of the parent animal." Scientists had already met the precedents, Steenstrup was saying, but they couldn't believe that they all belonged to the same species.

Steenstrup would eventually be proved right. Many parasites travel from one host to another during their life cycles, and in many cases they alternate between different forms from one generation to the next. And thanks to his insight, one of the best cases for spontaneous generation in parasites fell apart. Steenstrup turned his attention from flukes to the worms that Aristotle had seen living in cysts embedded in pig tongues. These parasites, called bladder worms at the time, can live in any muscle in mammals. Steenstrup suggested that bladder worms were actually an early stage in the development of some other worm not yet found.

Other scientists noticed that bladder worms looked a bit

like tapeworms. All you had to do was cut off most of the tape-worm's long ribbony body, and tuck its head and first few seg-ments inside a shell, and you had a bladder worm. Maybe the bladder worm and tapeworm were one and the same. Maybe they were actually the product of tapeworm eggs that had made their way into the wrong host. When the eggs hatched in this hostile environment, the tapeworms couldn't take their normal path of development but grew instead into stunted deformed monsters that died before they could reach maturity.

In the 1840s, a devout German doctor heard about these ideas and was outraged. Friedrich Küchenmeister kept a little medical practice in Dresden, and in his free time he wrote books on biblical zoology and ran the local cremation club, called Die Urne. Küchenmeister recognized that the idea that bladder worms were actually tapeworms certainly sidestepped the heresy of spontaneous generation. But it then fell into an-other sinful trap—the idea that God would let one of his crea-tures wind up in a monstrous dead end. "It would be contrary to the wise arrangement of Nature which undertakes nothing without a purpose," Küchenmeister declared. "Such a theory of error contradicts the wisdom of the Creator and the laws of harmony and simplicity put into Nature"—laws that even ap-plied to tapeworms.

Küchenmeister had a more pious explanation: the bladder worms were an early stage in the natural life cycle of the tape-worm. After all, the bladder worms tended to be found in prey—animals such as mice, pigs, and cows—and the tape-worms were found in predators: cats, dogs, humans. Perhaps when a predator ate prey, the bladder worm emerged from its cyst and grew into a full tapeworm. In 1851, Küchenmeister be-gan a series of experiments to rescue the bladder worm from its dead end. He plucked out forty of them from rabbit meat and fed them to foxes. After a few weeks, he found thirty-five tape-worms inside the foxes. He did the same with another species of tapeworm and bladder worm in mice and cats. In 1853, he fed bladder worms from a sick sheep to a dog, which soon was shedding the segments of an adult tapeworm in its feces. He fed these to a healthy sheep, which began to stumble sixteen days later. When the sheep was killed and Küchenmeister looked in its skull, he found bladder worms sitting on top of its brain.

When Küchenmeister reported his findings, he stunned the university professors who made parasites their life's work. Here was an amateur out on his own, sorting out a mystery the experts had failed to solve for decades. They tried to poke holes in Küchenmeister's work wherever possible, to try to keep their own ideas about dead-end bladder worms alive. One problem with Küchenmeister's work was that he sometimes fed the bladder worms to the wrong host species and the parasites all died. He knew, for example, that pork carried a species of bladder worm, and he knew that the butchers of Dresden and their families often suffered from tapeworms called *Taenia solium*. He suspected that the two parasites were one and the same. He fed *Taenia* eggs to pigs and got the bladder worms, but when he fed the bladder worms to dogs, he couldn't get adult *Taenia*. The only way to prove the cycle was to look inside its one true host—humans.

Küchenmeister was so determined to prove God's benevolent harmony that he set up a gruesome experiment. He got permission to feed bladder worms to a prisoner about to be executed, and in 1854 he was notified of a murderer to be decapitated in a few days. His wife happened to notice that the warm roast pork they were eating for dinner had a few bladder worms in it. Küchenmeister rushed to the restaurant where they had bought the pork. He begged for a pound of the raw meat, even though the pig had been slaughtered two days earlier and was beginning to go bad. The restaurant owners gave him some, and the next day Küchenmeister picked out the bladder worms and put them in a noodle soup cooled to body temperature.

The prisoner didn't know what he was eating and enjoyed it so much he asked for seconds. Küchenmeister gave him more soup, as well as blood sausage into which he had slipped bladder worms. Three days later the murderer was executed, and Küchenmeister searched his intestines. There he found young *Taenia* tapeworms. They were still only a quarter of an inch long, but they had already developed their distinctive double crown of twenty-two hooks.

Five years later, Küchenmeister repeated the experiment, this time feeding a convict four months before his execution. Afterward he found tapeworms as long as five feet in the man's intestines. He felt triumphant, but the scientists of his day were disgusted. The experiments were "debasing to our common nature," said one reviewer. Another compared him to some doctors of the day who cut the still-beating heart out of a just-executed man, merely to satisfy their curiosity. One quoted Wordsworth: "One that would peep and botanise/Upon his mother's grave?" But no doubt was left that parasites were among the strangest things alive. Parasites were not spontaneously generated; they arrived from other hosts. Küchenmeister also helped discover another important thing about parasites that Steenstrup hadn't observed: they didn't always have to wander through the outside world to get from one host to another. They could grow inside one animal and wait for it to be eaten by another.

The last possibility still left for spontaneous generation was represented by the microbes. That was shortly put to rest by the French scientist Louis Pasteur. To make his classic demonstration, he put broth in a flask. Given enough time the broth would go bad, filling with microbes. Some scientists claimed that the microbes were spontaneously generated in the broth itself, but Pasteur showed that the microbes were actually carried in the air to the flask and settled into it. He went on to prove that microbes weren't just a symptom of diseases but often their cause—what came to be known as the germ theory of infection. And out of that realization came the great triumphs of Western medicine. Pasteur and other scientists began to isolate the particular bacteria that caused diseases such as anthrax, tuberculosis, and cholera and to make vaccines for some of them. They proved that doctors spread disease with their dirty hands and scalpels and could stop it with some soap and hot water.

With Pasteur's work, a peculiar transformation came over the concept of the parasite. By 1900, bacteria were rarely called parasites anymore, even though, like tapeworms, they lived in and at the expense of another organism. It was less important to doctors that bacteria were organisms than that they had the power to cause diseases and that they could now be erased with vaccines, drugs, and good hygiene. Medical schools focused their students on infectious diseases, and generally on those caused by bacteria (or later, by the much smaller viruses). Part of their bias had to do with how scientists recognize causes of diseases. They generally follow a set of rules proposed by the German scientist Robert Koch. To begin with, a pathogen had to be shown to be associated with a particular disease. It also had to be isolated and grown in pure culture, the cultured organism had to be inoculated into a host and produce the disease again, and the organism in the second host had to be shown to be the same as that inoculated. Bacteria fit these rules without much trouble. But there were many other parasites that didn't.

Living alongside bacteria—in water, soil, and bodies—were much larger (but still microscopic) single-celled organisms known as protozoa. When Leeuwenhoek had looked at his own feces, he had seen a protozoan now called *Giardia lamblia*, which had made him sick in the first place. Protozoa are much more like the cells that make up our own bodies, or plants or fungi, than they are like bacteria. Bacteria are essentially bags of loose DNA and scattered proteins. But protozoa keep their DNA carefully coiled up on molecular spools within a shell called the nucleus, just as we do. They also have other compartments dedicated to generating energy, and their entire contents are surrounded by skeleton-like scaffolding, as with our cells. These were only a few of many clues biologists discovered that showed the protozoa to be more closely related to multicellular life than to the bacteria. They went so far as to divide life into two groups. There were the prokaryotes—the bacteria—and the eukaryotes: protozoa, animals, plants, and fungi.

Many protozoa, such as the amoebae grazing through forest floors, for instance, or the phytoplankton that turn the oceans green, are harmless. But there are thousands of species of parasitic protozoa, and they include some of the most vicious parasites of all. By the turn of the century, scientists had figured out that the brutal fevers of malaria weren't caused by bad air but by several species of a protozoan called *Plasmodium*, a parasite that lived inside mosquitoes and got into humans when the insects pierced the skin to suck blood. Tsetse flies carried trypanosomes that caused sleeping sickness. Yet, despite their power to cause disease, most protozoa couldn't live up to Koch's rigorous demands. They were creatures after Steenstrup's heart, passing through alternating generations.

Plasmodium, for example, enters a human body through a mosquito bite as a zucchini-shaped form known as a sporozoite. It travels to the liver, where it invades a cell and there multiplies into forty thousand offspring, called merozoites—these are now shaped like a grape. Merozoites pour out of the liver and seek out red blood cells, where they make more merozoites. The new generations burst out of the cells and seek out more blood cells. After a while, some of the merozoites produce a different form—a sexual one, called a macrogamont. If a mosquito should take a drink of the host's blood and swallow a blood cell with macrogamonts in it, they will mate inside the insect. The male macrogamont fertilizes the female one, and they produce a round little offspring called an ookinete. The ookinete divides in the mosquito's gut into thousands of sporozoites, which travel to the mosquito's salivary glands, there to be injected into some new human host.

With so many generations and so many different forms, you can't raise *Plasmodium* organisms simply by throwing them in a petri dish and hoping they'll multiply. You have to get male and female macrogamonts to believe that they're living in the gut of a mosquito, and once they've bred, you have to make their offspring believe they've been shot out of the mosquito's mouth and into human blood. It's not impossible to do, but it took until the 1970s, a century after Koch set up his rules, for a scientist to figure out how to culture *Plasmodium* in a lab.

Parasitic eukaryotes and parasitic bacteria were pushed further apart by geography. In Europe, bacteria and viruses caused the worst diseases, such as tuberculosis and polio. In the tropics, protozoa and parasitic animals were just as bad. The scientists who studied them were generally colonial physicians, and their specialty became known as tropical medicine. Europeans came to look upon parasites as robbing them of native labor, of slowing down the building of their canals and dams, of preventing the white race from living happily at the Equator. When Napoleon took his army to Egypt, the soldiers began to complain that they were menstruating like women. Actually they had been infected with flukes. Like the flukes Steenstrup had studied, these were shed by snails and swam through water looking for human skin. They ended up in the veins in the abdomens of the soldiers and pushed their eggs into their bladders. Blood flukes attacked people from the western shores of Africa to the rivers of Japan; the slave trade even brought them to the New World, where they thrived in Brazil and the Caribbean. The disease they caused, known as bilharzia or schistosomiasis, drained the energy of hundreds of millions of people who were supposed to build European empires.

As bacteria and viruses occupied the center of medicine, parasites (in other words, everything else) were spun out to the periphery. Specialists in tropical medicine went on struggling against their own parasites, often with a staggering lack of success. Vaccines against parasites failed miserably. There were a few old cures—quinine for malaria, antimony for blood flukes—but they did only a little good. Sometimes they were so toxic that they caused as much harm as the disease itself. Meanwhile, veterinarians studied the things living inside cows and dogs and other domesticated animals. Entomologists looked at the insects dug into trees, the nematodes that sucked on their roots. All these different disciplines became known as parasitology—more of a loose federation than an actual science. If anything held together its factions, it was that parasitologists were keenly aware of their subjects as living things rather than just agents of disease, each subject with a natural history of its own—in the words of one scientist at the time, "medical zoology."

Some actual zoologists studied this medical zoology. But just as the germ theory of disease was changing the world of medicine, they were reckoning with a revolution of their own. In 1859, Charles Darwin offered a new explanation for life. Life, he argued, hadn't existed unchanged since Earth's creation but had evolved from one form to another. That evolution had been driven by what he named natural selection. Every generation of a species was made up of variants, and some variants fared better than others—they could catch more food or avoid becoming food for someone else. Their descendants inherited their characteristics, and with the passing of thousands of generations, this unplanned breeding produced the diversity of life on Earth today. To Darwin, life was not a ladder rising up to the angels or a cabinet filled with shells and stuffed animals. It was a tree, bursting upward with all the diversity of the species on Earth alive today and long past, all rooted in a common ancestry.

Parasites fared as badly in the evolutionary revolution as they had in the medical one. Darwin contemplated them only in passing, usually when he was trying to argue that nature was a bad place to try to prove God's benevolent design. "It is derogatory that the Creator of countless systems of worlds should have created each of the myriads of creeping parasites," he once wrote. He found that parasitic wasps are a particularly good antidote to sentimental ideas about God. The way that the larvae devoured their host from the inside was so awful that Darwin once wrote of them, "I cannot persuade myself that a beneficient and omnipotent God would have designedly created the Ichneumonidae [one group of parasitic wasps] with the express intention of their feeding within the living bodies of Caterpillars."

Yet, Darwin was downright kind to parasites compared with the later generations of biologists who carried on his work. Instead of benign neglect, or even mild disgust, they felt outright scorn for parasites. These late Victorian scientists were drawn to a peculiar, now debunked form of evolution. They accepted the concept that life evolved, but Darwin's generation-by-generation filter of natural selection seemed too random to account for the trends they saw in the fossil record that had lasted millions of years. They saw life as having an inner force driving it toward greater and greater complexity. To their mind, this force brought a purpose to evolution: to produce the higher organisms—vertebrates such as us—from the lower beings.

One influential voice for these ideas belonged to the British zoologist Ray Lankester. Lankester grew up with evolution. When he was a boy, Darwin came to his family's house and told him stories about riding a giant tortoise on a Pacific island. When Lankester became a man, he had a giant frame and a puffy, vaguely Charles Laughton–like face. As an Oxford professor and the director of the British Museum he carried Darwin's theory forward with what seemed at times like sheer bodily power. He made the people around him feel small in both size and mind; he reminded one man who met him of a winged Assyrian beast. Once King Edward VII offered him some tidbit of scientific knowledge while paying him a royal visit, and Lankester bluntly replied, "Sir, the facts are not so; you have been misinformed."

To Lankester, Darwin's theory had brought a unity to biology as impressive as that in any other science. He had no patience for doddering dons who looked at his science as a quaint hobby. "We are no longer content to see biology scoffed at as inexact or gently dropped as natural history or praised for her relation to medicine. On the contrary, biology is the science whose development belongs to the day," he declared. And its understanding would help free future generations from stupid orthodoxies of all sorts: "the jack-in-office, the pompous official, the petulant commander, the ignorant pedagogue." It would help carry human civilization upward, as life itself had been striving for millions of years. He laid out this view of the biological and political order of things in an essay he wrote in 1879, titled "Degeneration: A Chapter in Darwinism."

The tree of life you find described in that essay isn't the wild bush of Darwin. It's shaped like a plastic Christmas tree, with branches sticking out to the side from a main shaft, which rises to higher and higher glories until it reaches humans at the top. At each stage in the rise of life, some species abandoned the struggle, comfortable with the level of complexity they had achieved—a mere amoeba, sponge, or worm—while others kept striving upward.

But there were some drooping branches on Lankester's tree. Some species not only stopped rising but actually surrendered some of their accomplishments. They *degenerated*, their bodies simplifying as they accommodated themselves to an easier life. For biologists of Lankester's day, parasites were the sine qua non of degenerates, whether they were animals or single-celled protozoa that had given up a free life. To Lankester, the quintessential parasite was a miserable barnacle named *Sacculina carcini*. When it first hatched from its egg, it had a head, a mouth, a tail, a body divided into segments, and legs, which is exactly what you'd expect from a barnacle or any other crustacean. But rather than growing into an animal that searched and struggled for its own food, *Sacculina* instead found itself a crab and wiggled into its shell. Once inside, *Sacculina* quickly degenerated, losing its segments, its legs, its tail, even its mouth. Instead, it grew a set of rootlike tendrils, which spread throughout the crab's body. It then used these roots to absorb food from the crab's body, having degenerated to the state of a mere plant. "Let the parasitic life once be secured," Lankester warned, "and away go legs, jaws, eyes, and ears; the active, highly gifted crab may become a mere sac, absorbing nourishment and laying eggs."

Since there was no divide between the ascent of life and the history of civilization, Lankester saw in parasites a grave warning for humans. Parasites degenerated "just as an active healthy man sometimes degenerates when he becomes suddenly possessed of a fortune; or as Rome degenerated when possessed of the riches of the ancient world. The habit of parasitism clearly acts upon animal organization in this way." To Lankester, the Maya, living in the shadows of the abandoned temples of their ancestors, were degenerates, just as Victorian Europeans were pale imitations of the glorious ancient Greeks. "Possibly we are all drifting," he fretted, "tending to the condition of intellectual Barnacles."

An uninterrupted flow from nature to civilization meant that biology and morality were interchangeable. People of Lankester's day took to condemning nature and then using nature in turn as an authority to condemn other people. His essay inspired a writer named Henry Drummond to publish a best-selling screed, *Natural Law in the Spiritual World*, in 1883. Drummond declared that parasitism "is one of the gravest crimes in nature. It is a breach of the law of Evolution. Thou shalt evolve, thou shalt develop all thy faculties to the full, thou shalt attain to the highest conceivable perfection of thy race—and so perfect thy race—this is the first and greatest commandment of Nature. But the parasite has no thought for its race, or for its perfection in any shape or form. It wants two things—food and shelter. How it gets them is of no moment. Each member lives exclusively on its own account, an isolated, indolent, selfish, and backsliding life." People were no different:

"All those individuals who have secured a hasty wealth by the chances of speculation; all children of fortune; all victims of inheritance; all social sponges; all satellites of the court; all beggards of the market-place—all these are living and unlying witness to the unalterable retributions of the law of parasitism."

People had been referred to as parasites before the late 1800s, but Lankester and other scientists gave the metaphor a precision, a transparency, that it never had before. And it's a short walk from Drummond's rhetoric to genocide. Listen to how closely his line about the highest conceivable perfection of a race meshes with these words: "In the struggle for daily bread all those who are weak and sickly or less determined succumb, while the struggle of the males for the females grants the right or opportunity to propagate only to the healthiest. And struggle is always a means for improving a species' health and power of resistance and therefore, a cause of its higher development." The author of these words wasn't an evolutionary biologist but a petty Austrian politician who would go on to exterminate six million Jews.

Adolf Hitler relied on a confused, third-rate version of evolution. He imagined that Jews and other "degenerate" races were parasites, and he took the metaphor even further, seeing them as a threat to the health of their host, the Aryan race. It was the function of a nation to preserve the evolutionary health of its race, and so it had to rid the parasite from its host. Hitler probed every hidden turn of the parasite metaphor. He charted the course of the Jewish "infestation," as it spread to labor unions, the stock exchange, the economy, and cultural life. The Jew, he claimed, was "only and always a parasite in the body of other peoples. That he sometimes left his previous living space has nothing to do with his own purpose, but results from the fact that from time to time he was thrown out by the host nations he had misused. His spreading is a typical phenomenon for all parasites; he always seeks a new feeding ground for his race."

Nature's Criminals

Nazis weren't the only ones to burn the brand of parasite on their enemies. To Marx and Lenin, the bourgeoisie and the bureaucrats were parasites that society had to get rid of. An exquisitely biological take on socialism appeared in 1898, when a pamphleteer named John Brown wrote a book called *Parasitic Wealth or Money Reform: A Manifesto to the People of the United States and to the Workers of the World.* He complained of how three-quarters of the country's money was concentrated in the hands of 3 percent of the population, that the rich sucked the wealth of the nation away, that their protected industries flourished at the people's expense. And, like Drummond or Hitler, he saw his enemies precisely reflected in nature, in the way parasitic wasps live in caterpillars. "With the refinement of innate cruelty," he wrote, "these parasites eat their way into the living substance of their unwilling but helpless host, avoiding all the vital parts to prolong the agony of a lingering death."

Parasitologists themselves sometimes helped consecrate the human parasite. As late as 1955, a leading American parasitologist, Horace Stunkard, was carrying on Lankester's conceit in an essay published in the journal *Science*, titled "Freedom, bondage, and the welfare state." "Since zoology is concerned with the facts and principles of animal life, information obtained from the study of other animals is applicable to the human species," he wrote. All animals were driven by the need for food, shelter, and the chance to reproduce. In many cases, fear drove them to give up their freedom for some measure of security, only to be trapped in permanent dependency. Conspicuous among security-seeking animals were creatures such as clams, corals, and sea squirts, which anchored themselves to the ocean floor in order to filter the passing sea water for food. But none could compare with the parasites. Time after time in the history of life, free-living organisms had surrendered their liberty to become parasites in exchange for an escape from the dangers of life. Evolution then took them down a degenerate path. "When other food sources were insufficient, what would be

19

easier than to feed upon the tissues of the host? The dependent animal is proverbially looking for the easy way."

Stunkard was only a little coy about how this rule of parasites could apply to humans. "It may be applied to any group of organisms, and is not intended to refer merely to political entities, although certain implications may be in order." With its complete surrender of its liberty, the parasite had entered the "welfare state," as Stunkard put it—with hardly a tissue of metaphor dividing the tapeworm and the New Deal. Once parasites gave up their freedom, they rarely managed to regain it; instead, they channeled their energies into making new generations of parasites. Their only innovations were weird kinds of reproduction. Flukes alternated their forms between generations, reproducing sexually in humans and asexually in snails. Tapeworms could produced a million eggs a day. How could Stunkard have had anything but fast-breeding welfare families in mind? "Such a welfare state exists only for those lucky individuals, the favored few, who are able to cajole or compel others to provide the welfare," he wrote. "The well-worn attempt to obtain comfort without effort, to get something for nothing, persists as one of the illusions that in all ages has intrigued and misled the unwary."

Writing in 1955, Stunkard represented a dying gasp of the old take on evolution. As he was attacking food-stamp parasites, his fellow biologists were unceremoniously dumping the whole foundation of his scientific view. They discovered that every living thing on Earth carries genetic information in its cells in the form of DNA, a molecule in the shape of a double helix. Genes (particular stretches of DNA) carried the instructions for making proteins, and these proteins could build eyes, digest food, regulate the creation of other proteins, and do thousands of other things. Each generation passed its DNA to the next, and along the way the genes got shuffled into new combinations. Sometimes mutations to the genes turned up, creating new codes altogether. Evolution, these biologists realized, was built on these genes and the way they rose and fell as time passed—not on some mysterious inner force. The genes offered up rich variety, and natural selection preserved certain kinds. From these genetic ebbs and flows new species could be created, new body plans. And since evolution was grounded on the short-term effects of natural selection, biologists no longer had any need for an inner drive for evolution, no longer saw life as a plastic Christmas tree.

Parasites should have benefited from this change of scientific heart. They were no longer the backward pariahs of biology. Yet, well into the twentieth century, parasites still couldn't escape Lankester's stigma. The contempt survived both in science and beyond it. Hitler's racial myths have collapsed, and the only people who still believe in eradicating social parasites are at the fringes, among the Aryan skinheads and the minor dictators. Yet, the word *parasite* still carries the same insulting charge. Likewise, for much of the twentieth century, biologists thought of parasites as minor degenerates, mildly amusing but insignificant to the pageant of life. When ecologists looked at how the sun's energy streamed through plants and into animals, parasites were nothing more than grotesque footnotes. What little evolution parasites experienced was the result of being dragged along by their hosts.

Even in 1989, Konrad Lorenz, the great pioneer in animal behavior, was writing about the "retrograde evolution" of parasites. He didn't want to call it *degeneration*—that word was perhaps too loaded by Nazi rhetoric—and so he replaced it with "sacculinasation," after *Sacculina*, Lankester's backsliding barnacle. "When we use the terms 'higher and lower' in reference to living creatures and to cultures alike," he wrote, "our evaluation refers directly to the amount of information, of knowledge, conscious or unconscious, inherent in these living systems." And according to this scale, Lorenz despised parasites: "If one judges the adapted forms of the parasites according to the amounts of retrogressed information, one finds a loss of information that coincides with and completely confirms the low estimation we have of them and how we feel about them. The mature *Sacculina carcini* has no information about any of the particularities and singularities of its habitat; the only thing it knows anything about is its host." Much like Lankester 110 years earlier, Lorenz saw the only virtue of parasites as a warning to humans. "A retrogression of specific human characteristics and capacities conjures up the terrifying specter of the less than human, even of the inhuman."

From Lankester to Lorenz, scientists have gotten it wrong. Parasites are complex, highly adapted creatures that are at the heart of the story of life. If there hadn't been such high walls dividing scientists who study life—the zoologists, the immunologists, the mathematical biologists, the ecologists—parasites might have been recognized sooner as not disgusting, or at least not merely disgusting. If parasites were so feeble, so lazy, how was it that they could manage to live inside every free-living species and infect billions of people? How could they change with time so that medicines that could once treat them became useless? How could parasites defy vaccines, which could corral brutal killers like smallpox and polio?

The problem comes down to the fact that scientists at the beginning of this century thought they had everything figured out. They knew how diseases were caused and how to treat some of them; they knew how life evolved. They didn't respect the depth of their ignorance. They should have borne in mind the words of Steenstrup, the biologist who had first shown that parasites were unlike anything else on Earth. Steenstrup had it right in 1845 when he wrote, "I believe that I have given only the first rough outlines of a province of a great terra incognita which lies unexplored before us and the exploration of which promises a return such as we can at present scarcely appreciate."

2

Terra Incognita

May I never lose you, oh, my generous host, oh, my universe. Just as the air you breathe, and the light you enjoy are for you, so you are for me.

—Primo Levi, *Man's Friend*

Raquel Welch would have fared pretty poorly without her submarine. Suppose she had been shrunk down to the size of a pinhead and then had to get into the bloodstream of the dying diplomat on her own. Even if she could have clawed her way through the tough layers of skin and wiggled into a blood vessel, she would then have been sent flailing through his circulatory system by the pulsing push of his heart. Let's say, for the sake of argument, that she could wear a scuba-like mask that could pull the oxygen out of the blood so she could breathe. She'd still suffocate if she ended up in some part of the body where there's hardly any oxygen at all, like the liver. And as she tumbled through the darkness she'd be utterly lost, with no idea whether she was in the vena cava or the carotid artery.

The inside of a body is a tough place to survive. With our air-breathing lungs, our ears finely tuned to the vibrations of the air, we are adapted to life on land. A shark is made for the sea, ramming water through its gills and smelling for prey miles away. Parasites live in a different habitat altogether, one for which they are precisely adapted in ways that scientists only barely understand. Parasites can navigate through their murky labyrinth; they can glide through skin and gristle; they can pass unscathed through the cauldron of the stomach. They can turn just about every organ in the body—the eustachian tube, the gill, the brain, the bladder, the Achilles tendon—into their home. They can rebuild parts of the host's body to suit their own comfort. They can feed on almost anything: blood, gut lining, liver, snot. They can make their host's body bring them food.

Parasitologists need years, sometimes decades, to decipher these adaptations. They can't spend a summer following a troop of monkeys or put radio collars on a pack of wolves. Parasites live invisibly, and parasitologists usually can see what they're doing only by killing their hosts and dissecting them. These grisly snapshots slowly add up to a natural history.

Steenstrup knew that flukes were extraordinary animals, but little more than that. After one hundred fifty years of experiments, parasitologists can show just how extraordinary they are. Consider the blood fluke *Schistosoma mansoni*, a tiny missile just emerged from its snail and swimming through a pond in search of a human ankle. If it feels the ultraviolet rays of the sun, it stops swimming and sinks back down into the darkness to hide from the damaging radiation. But if it senses molecules from human skin, it begins to swim madly, jerking around in different directions. When it reaches the skin, it drills its way in. Human skin is far tougher than the soft flesh of a snail, so the fluke lets its long tail snap off, the wound quickly healing as it burrows in. Special chemicals it releases from its coat soften up the skin, letting it plunge into its host like a worm in mud.

After a few hours it has reached a capillary. It has traded the streams of the outside world for the internal ones. These capillaries are barely wider than the fluke itself, so the fluke needs to use a pair of suckers to inch forward. It makes its way to a larger vein, and a larger one still, finally making its way into a torrent of blood so powerful it carries the fluke away. The parasite rides the surge until it finally reaches the lungs. It moves from the veins to the arteries like a snake in a forest canopy. Finding its way back into a lung capillary, and then to a major artery, it is swept through the body once again. It may tour its host's entire body three times until it finally comes to rest in the liver.

Here the fluke lodges itself in a vessel and finally has its first meal since leaving the snail: a drop of blood. It now begins to mature. If it's a female, a uterus starts to take shape. If it's a male, eight testes form like a bunch of grapes. In either case, the fluke grows dozens of times bigger in a few weeks. Now it is time for the parasite to search for a partner for life. If it is lucky, other flukes sniffed out this human host and are lodged in the liver as well. The females are delicate and slender; the males are shaped something like a canoe. They begin to make blood-borne odors that lure members of the opposite sex, and once a female encounters a male, she slips into his spiny trough. There she locks in, and the male carries her out of the liver. Over the course of a couple of weeks, the pair make the long journey from the liver to the veins that fan out across the gut. As they travel the male passes molecules into the female's body that tell her genes to make her sexually mature. They keep traveling until they reach a resting place unique to their own species. *Schistosoma mansoni* stops near the large intestine. If we were following *Schistosoma haemotobium*, it would take another route to the bladder. If we were following *Schistosoma nasale*, a blood fluke of cows, it would take yet another route to the nose.

Once they find their destined place, the fluke couple stay

there for the rest of their lives. The male drinks blood with his powerful throat and massages the female to help thousands of blood cells flow into her mouth and through her gut; he consumes his own weight in glucose every five hours and passes on most of it to her. They may be the most monogamous couples in the animal kingdom—a male will clasp onto its female even after she has died. (A few homosexual flukes will also get together. While their fit isn't as tight, they will keep reuniting if a disapproving scientist should separate them.)

Heterosexual flukes mate every day of their long lives, and whenever the female is ready to lay her eggs, the male makes his way along the wall of the bowels until he finds a good spot. The female slides partially out of the trough, far enough to lay her eggs in the smallest capillaries. Some of the eggs are carried away by the bloodstream and end up back in the liver, that meaty filter, where they lodge and inflame the tissue, causing much of the agony of schistosomiasis. But the rest of the eggs work their way into the intestines and escape their host, ready to slice open their shells and find a new snail.

Each piece of the parasite puzzle costs years of research. The question of how parasites navigate has taken up just about the entire career of one scientist, Michael Sukhdeo. Sukhdeo teaches these days at Rutgers University in New Jersey. New Jersey may be a long way from Tambura, but he has no shortage of parasites to study in horses, cows, and sheep. I paid a visit to Sukhdeo at his office. He is a stocky man with a sly goatee. A bike hangs from his office wall, fish swim in a tank by his desk, and classic rock comes out of his radio. Sukhdeo, like a lot of parasitologists I've met, can slide into gruesome conversation without any warning. I suppose when you spend your days studying creatures that chew up the lining of livers and intestines, there's no sense in dancing around the uglier basics of life. He started to talk about how grotesque it is when people get elephantiasis, which was common in British Guyana, where he spent much of his childhood. "Everywhere you walked you saw people with huge bulges in their crotch and big swollen elephantine feet," he said.

Sukhdeo then told me how he himself became infected when he was eleven. He developed a swelling, and his parents took him to a clinic. "When you're testing for elephantiasis, the microfilaria come out into the bloodstream only at dusk. Nobody knows where they go. So at night we had to go to this clinic to get our blood checked. And there was a girl there, about my age; she was eleven, and she had only one breast. That's a place where the worms live. She was a beautiful girl; I was in love. We both got checked at the same time. It was twelve Guyanese dollars—six American dollars—for treatment. They couldn't afford it for their daughter. We offered to pay for them, but they were very proud and wouldn't even take a loan. And so that girl remained infective—over six American dollars."

Sukhdeo went to McGill University in Montreal, and there he discovered that while parasites might be grotesque, they were also the most interesting creatures he had ever encountered. "I took a course in human parasitology, and—pow—it was disgusting and really exciting. I had gone through four years of university and nothing had turned me on in just that way. They were just so weird, and there was so little known about them."

He decided to go on studying parasites in graduate school, and there he realized that people had very little idea of how parasites behaved as actual, living organisms. Many parasitologists have resigned themselves to studying them on an abstract plane—cataloging new species according to their suckers and spines, for example, without ever knowing what those suckers and spines are for.

For his master's degree, Sukhdeo chose *Trichinella spiralis*. This tiny nematode comes our way inside the muscle of undercooked pork, where it lives in cysts formed from individual muscle cells. When a person eats the meat, the parasite breaks out of its cyst and makes its way into the intestines, threading itself through the cells of the lining. There it mates and produces a new generation of *Trichinella*, which leave the intestines and travel through the bloodstream until they lodge in the person's muscle and form cysts of their own. Humans are only accidental hosts for *Trichinella*; they are unable to carry the parasite on to the next stage of its life cycle. Pigs are a much more profitable host; a dead pig may be scavenged by a rat, which then dies and is scavenged by another rat, which may be eaten by a pig. Pigs can pass *Trichinella* to each other by being fed infected meat or by chewing their tails off. In the wild, predatory mammals and scavengers keep the cycle spinning along—ranging from polar bears and walrus in the Arctic to hyenas and lions in Africa.

The parasites traveling each of these cycles had been designated as individual species, but no one actually knew whether they weren't actually a single species scattered among different regions and hosts. Sukhdeo got hold of *Trichinella* from Russia, from Canada, and from Africa, as he was told, and he ground up each sample and infected mice with them. He extracted the antibodies that the mice's immune systems produced against the ground-up parasites and compared them to judge how similar they were to one another.

Eventually he stopped to wonder why he was doing what he was doing. His experiments were based on the assumption that individuals of a species look similar to one another. This is usually a pretty reliable assumption, but biologists have recognized that it's not always the case. Poodles and Dobermans belong to the same species, for instance. On the other hand, two beetles that look practically identical may belong to separate species. Rather than focus on appearances, biologists these days define a species as a group of organisms that breed together and don't breed with other groups. It's out of that isolation that evolution then makes a species distinct from others.

Sukhdeo decided that the best way to study the species of his parasites was to work out their sex life. He dissected

Trichinella cysts out of muscle and teased out the worms, only 250 microns long. He'd check their sex and then get the parasite into a syringe, which he'd inject into the stomach of a mouse. Then he'd go back to his cysts and find a parasite of the opposite sex, and then inject it into the mouse's stomach as well. A month later he'd look at the mouse's muscle to see whether they had mated and produced young.

Sukhdeo concluded that the African form was probably a subspecies and not a separate species of its own. But the experiment actually raised a much deeper, much more interesting question. How did the parasites find each other?

Apply the *Fantastic Voyage* method: It would be as if you were thrown down into a dark cavernous tunnel twelve miles long, lined on all sides with slippery, tightly packed, man-sized mushrooms. If you were set down randomly in there and moved around randomly, there'd be no hope of finding someone else in such a place. And yet, *Trichinella*—without a map or even much of a brain—always did.

Sukhdeo wanted to know how they did it, but his adviser told him not to try. "'You can't find out how these animals go wherever they go because for a hundred years parasitologists have been trying to find out the answer and they haven't been able to. Better people than you have tried.'"

Sukhdeo ignored the advice and set out to find the secret to parasite navigation. Unfortunately, he set out in the wrong direction. He assumed that like animals on the outside, parasites must follow a gradient. A shark smells the blood of a wounded seal from miles away and heads for it, thanks not only to its sharp nose but to the simple law of how blood spreads in water. The farther away the blood travels from the seal, the thinner it gets. If a shark keeps heading along a rising gradient, it will automatically reach the source. As soon as it veers away in the wrong direction, the blood trails off, and it can right itself. Gradients work in the air just as well as in the water. They help lead bees to flowers and hyenas to carcasses. Tracking gradients works so well at sea and on land that it only made sense that parasites must use them as well. Parasitologists searched for the scent of a gallbladder, the whiff of an eye. They didn't find any.

For years, Sukhdeo tried to find the secret for himself. He built chambers out of Plexiglas in which he could put a parasite, and then he'd add different chemicals to see if it would swim toward it. At first he kept his entire lab heated to body temperature. Then he invented a system of tubes to circulate warm water around his artificial gut. "I would try to sample everything they encountered in the host. First I tried salivary secretions, and then I would move down the gut." Nothing he did made sense. He couldn't get the parasites to swim toward or away from any substance he put in the chamber.

They did react sometimes, but in a way that made no sense at all. "Whenever these little parasites encountered bile they started moving like crazy," Sukhdeo said. "That wasn't what I wanted—I wanted something that attracted them. Initially they would move back and forth fifty times a minute, and if you put bile in, there was an instantaneous change and they started moving sinusoidally."

Sukhdeo kept looking for the key to parasite navigation after he moved to the University of Toronto. As he searched he drifted into an academic limbo. At Toronto he met his wife, Suzanne, who was also getting her Ph.D. in parasitology with the director of his lab. When the director developed Alzheimer's disease, Sukhdeo took over the lab and became Suzanne's dissertation advisor. If he had wanted to have a real career in parasitology, he should have been looking for jobs elsewhere, but instead he lingered in Toronto, applying for more money each year to carry on his experiments. For six years he floated in this dead-end existence, but he found that it gave him the freedom to search for answers that other scientists thought were unreachable. "I had nothing to lose," Sukhdeo says. "I could do anything I wanted, and I had no future."

He decided to extend his research to other species, such as

the liver fluke, *Fasciola hepatica*. A relative of the blood fluke, it has a similar life cycle. It lives inside cows and other grazing mammals, and its eggs pass out of its host's body with feces. It hatches from its egg and swims in search of a snail, where a couple of generations grow up. Cercariae emerge from the snail and swim away from the snail until they hit any object—usually a rock or a plant—and build themselves a tough transparent cyst. When another grazing mammal eats them, their acid-proof shell carries them safely through the stomach and into the intestines. Once in the intestines, they break loose and burrow out into the abdominal cavity and then head for the liver. There they grow into adults—leaf-shaped inch-long animals that can cram into a liver by the hundreds and live for eleven years. Liver flukes can sometimes get into humans, but the real danger they pose is to livestock. In tropical countries, between 30 and 90 percent of cattle carry them, and they cause $2 billion in damage every year. Yet, despite the massive harm they cause and despite decades of research, scientists had no idea how they managed to find the liver.

Sukhdeo built himself new chambers out of brass and aluminum and put liver flukes into them. He spent three years trying out different compounds given off by the liver—chemicals that might lure the flukes to their final home. Out of sheer exasperation, he tracked down a prominent liver physiologist to see if there was some attractant he had overlooked.

"He thought about it for a long time and said, 'You know, son, around the liver there is a capsule; it's called Glisson's capsule?'"

"'I said, 'Yes.'

"He said, 'Well, that's the end of my universe.'"

Sukhdeo found that while he couldn't get liver flukes to swim upstream to any particular cue, certain chemicals like bile made them react violently. He had seen the same strange reaction in *Trichinella* when he exposed it to the chemical pepsin. And then, as he was chewing over his data he realized that he had been looking at the problem from the wrong angle all along. He had been looking at the fluke or worm as a free-living creature, not as a parasite. A body is not a peaceful ocean. It's a sealed space in which fluids churn and slosh. A scent released from one organ can't spread smoothly and tranquilly through other organs. An airborne odor spreads out evenly, essentially to infinity, but a chemical marker inside a body must come up against any of a number of barriers, bouncing back and saturating the territory, destroying any clues it might have offered.

Sukhdeo explained his realization to me in his office, waving his arms at the wall. "For a gradient to form, you need an open-ended system, and you can't have turbulence. If I put a piece of toast here, you would smell it and know where it is. If I closed the room, quickly it would saturate. Because it's in a closed sysem, you can't have a gradient. If you put guts in this room, they would do the same thing."

The world of a parasite isn't like our own—it has its own constraints and opportunities. Because of the strange conditions found inside a body, Sukhdeo wondered whether parasites might be able to navigate not with gradients but by simply reacting to a few different sorts of stimuli. Konrad Lorenz had shown that free-living animals in the outside world rely on reflexive behaviors when they find themselves in predictable situations. If you're a goose and one of your eggs starts to roll out of your nest, you can perform a set of automatic actions to get it back: *stick out neck, pull back neck, bend head down.* That should get the egg under your beak and back into the nest without requiring you to pay much attention to the egg itself. If a biologist should sneak a goose's egg out from under its beak in the middle of this sequence, the goose will keep pulling its neck back anyway.

Sukhdeo wondered whether parasites relied on these kinds of programmed behavior more than free-living creatures. A body is in some ways more predictable than the outside world. A mountain lion born in the Rockies has to learn the shape of its territory and relearn it whenever a fire or a landslide or a parking lot suddenly changes the topography. A parasite can travel through a rat, safe in the knowledge that it crawls through a little biosphere that's almost identical with any other rat interior. The heart is always between the lungs, the eyes in front of the brain. By reacting in a certain way to certain landmarks on their journey, parasites can be transported where they need to go. "Everything else is irrelevant," says Sukhdeo. "They don't have to waste time generating neurons to recognize everything else that's going on."

Now all the weird behavior of *Trichinella* and liver flukes settled down into straightforward recipes for success. *Trichinella* sits tight in its muscle capsule as it falls into the stomach. There it picks up one of the chemicals, known as pepsin, that breaks food down in the stomach; in response, *Trichinella* starts to flail. "The first movement causes them to break out of that cyst. You can see them whipping until the tail lashes out and they're out in the stomach." The piece of meat they're lodged inside passes out of the stomach and into the intestines, where there's a duct from the liver down which bile flows to help with digestion. And bile is the second trigger, making them change from their whipping movement to a snakelike slither. That lets them move out of the food and into the intestines.

Sukhdeo figured out a way to test this idea. "What if I changed where the bile came in?" he said. "I had learned a lot about surgery, and I could stick a cannula with bile anywhere I wanted." Wherever along the intestines he moved the source of the bile was where *Trichinella* would settle. "The only reason they went where they went was because of bile."

Sukhdeo turned to his liver flukes, and he found that they also followed rules instead of gradients. Because they have a longer journey than *Trichinella*, they need three rules instead of two. When a liver cyst tumbles into the intestines, it's sensitive to bile as well. When it senses it, it starts twitching—"it goes spastic," says Sukhdeo. As it writhes, it breaks open its cyst, and the same movements drive it through the mushy wall of the intestines and into the abdominal cavity. A liver fluke has two suckers, one by its mouth and one by its belly. It can crawl by extending its front sucker, clamping it down, and then pulling up the rest of its body and anchoring it with the belly sucker. Flukes can also crimp—their whole body suddenly contracts in a violent spasm, and they let go of both suckers.

These kinds of movements are all that a fluke needs to get to the liver. It doesn't need a copy of *Gray's Anatomy* showing it the way. When it emerges out of the small intestines, it crimps itself out into the abdominal cavity, eventually reaching the smooth wall of abdominal muscles. The following day, the fluke switches to creeping. Now safe from the torrents of the intestines, it creeps along the abdominal wall without having to worry about getting washed away.

At this point, a creeping liver fluke will almost always reach the liver, no matter which way it travels. You might expect that the fluke at least has to know a few things: which way is up and which is down, for example, or the fact that the liver is next to the pancreas but not the gallbladder. Not so. The fluke takes advantage of the fact that the abdominal cavity is like the inside of a beach ball. Even if it crawls straight down to the bottom, it will reach the liver if it simply continues to crawl in a straight line, coming back around to the top, where the liver sits. That's why Sukhdeo found that 95 percent of flukes enter the liver from its upper side where it meets the diaphragm—the summit of the abdominal cavity. Despite the fact that a liver's underside is big and closer to the intestines, only 5 percent penetrate it from that side.

It took a decade for Sukhdeo to figure out how these two parasites navigate. These days he is almost respectable. To his surprise, he was offered a job as a parasitologist at Rutgers despite his years in limbo. He has a lab full of students eager to decipher the navigation of other parasites. He's thinking of ways to turn his discoveries into a way to kill parasites by giving them navigation signals at the wrong time. And he has many more puzzles to work on. When I last spoke to Sukhdeo, he was working on another fluke. It also starts out in a snail, but when it emerges from this host, it seeks out a fish instead of a sheep. As the fish swims past, the fluke snags onto the fish's tail and burrows into the meat. It then makes a beeline through the muscle for the fish's head and comes to rest within the lens of the fish's eye. "It seems that all the ideas people had before were wrong, so we're starting from scratch," he said.

Sukhdeo has earned the respect of other parasitologists for having shown that there is a behavior to parasites, that they make their way through the unique inner ecology of their hosts' bodies, and that you can figure out the rules they obey. He even got an award not long ago for his work, a plaque that he hands to visitors with a puzzled look. "When they gave it to me, I said, 'Why am I getting this?' I had been blackballed for so many years." There's a note of nostalgia when he talks about being ignored and ridiculed. He once submitted a paper to a journal about animal behavior and was rejected. When he asked the editor why, the editor reread the paper and accepted it, saying, "I had no idea parasites behaved. Please excuse my vertebrate chauvinism." And his old advisor wasn't the only parasitologist to tell him he was making a mistake. "At a meeting I went to, I was saying that we had to use ecological concepts when we were looking at parasites, and I got this old parasitologist standing up and shouting 'Heresy!' with the spittle coming up. A heretic!"

The word made Sukhdeo smile, and at that moment his goatee looked particularly devilish. "It was the high point of my career."

●　●　●

Once a parasite manages to find the place in its host where it will live, it can't just sit back and enjoy life. For one thing, it needs a way to stay put in its new home. As an adult, a liver fluke is adapted only to life in the liver; put it in the heart or the lung and it will die. For every place that parasites have to live, evolution has produced a way for them to stay there. For example, there are parasitic copepods (a kind of crustacean) that live all over the bodies of fish. There are copepods that live in the eye of the Greenland shark. There are copepods that live on the scales of Mako sharks, and others that live on their gill arches. There are copepods that live inside the noses of blue sharks. There are copepods that ram themselves through the side of a swordfish and clamp onto its heart.

Each of these copepods looks so different from the other species that it's hard for anyone except an expert to see that they all evolved from a common ancestor. Far from degenerating, these copepods have developed into bizarre forms in order to hold tight in their chosen niches. If these copepods should lose their grip, they would float away to a certain death. Every shark has its own special geometry of its scales, and copepods that live on the scales clasp their legs around them perfectly, like a lock and key. The copepod that lives in the Greenland shark has turned one of its legs into a mushroom-shaped anchor that it rams into the jelly of the eye.

Even for tapeworms, snug in the intestine, staying in place takes major effort. As they feed, tapeworms grow at a spectacular rate, increasing their size by a factor of as much as 1.8 million in two weeks. They can't eat the way most animals do, because they have no mouth or gut. Their digestion doesn't happen on the inside of their body but rather on the outside, their skin consisting of millions of delicate, blood-filled, fingery projections that can absorb food. The intestines of their host are also lined with almost identical projections. You could say that a tapeworm isn't really missing a digestive tract—it's an intestine turned inside out.

Tapeworms live in surging tides of half-digested food, blood, and bile, driven by the intestine's endless peristalsis. If they do nothing, peristalsis will carry tapeworms out of their host altogether. Some species of tapeworms clamp themselves to the intestines with hooks and suckers on their heads, but others are perpetually slithering to where the food is. When we eat, peristalsis immediately ripples through our intestines, and these unanchored tapeworms respond by swimming upstream. They reach the incoming food and keep swimming until they hit the highest concentration. At that point, they soak up their meal through their skin, but as they eat, the food is carried downstream, and for a while the tapeworms let themselves be carried along with their movable feast. All the while, the tapeworms keep track of how far they've drifted by sensing how their host's peristalsis changes. If they move too far downstream, they stop eating and swim back up. As tapeworms grow to their spectacular lengths, this swimming upstream can get to be complicated. The trouble is that peristalsis may make the intestines ripple quickly in one place and not at all farther up. Somehow tapeworms can detect these differences. They respond by making some parts of their body swim fast and some slowly.

The intestines are also home to hookworms, parasites that play a far riskier game whenever they eat. Hookworms start their lives in damp soil, where they hatch from eggs and grow into tiny larvae. They can travel into a human body by two routes: one simple, one tortuous. If a person swallows a larva, it will travel straight down to the intestines. But hookworms, like blood flukes, can penetrate the skin and burrow into a capillary. They swim through the veins to the heart and the lungs. When their host coughs, the larvae are carried up into their throat and can head down the esophagus.

Once it gets into the intestines, the hookworm grows into an adult, about half an inch long. Unlike tapeworms, the hookworm has a mouth—a powerful one ringed with daggerlike teeth and attached to a powerful, muscle-lined esophagus. And unlike tapeworms, it's not interested in the half-digested food flowing through the intestines but in the intestines themselves. It drives it mouth into the lining of the intestines, ripping up the flesh. Parasitologists are still debating whether hookworms then drink their host's blood or sop up the torn-up intestinal tissue. In either case, they release their grip after a while and swim to a new patch of tissue to feed.

But when the hookworm tears up some intestine and puts it in its mouth, the blood starts to clot. Whenever a blood vessel is torn, it picks up molecules from the cells in the surrounding tissue. Some of these new molecules combine with compounds floating in the blood itself. These chemicals trigger a cascade of reactions with other factors in the blood, which ultimately activate special cells known as platelets. The platelets swarm to the wound and clump together, while the cascade also creates a mesh of fibers around them, forming a hard clot that stops the bleeding. For a hookworm, clotting can mean starvation as the blood vessels in its mouth turn hard.

The parasite responds with a sophistication biotechnologists can only ape. It releases molecules of its own that are precisely shaped to combine with different factors in the clotting cascade. By neutralizing them, the hookworm keeps the platelets from clumping and allows the blood to keep flowing into its mouth. Once a hookworm finishes feeding at one place, the vessels can recover and clot while the parasite moves on to a fresh bit of intestines. If the hookworm were to use some crude blood-thinner that flooded the intestines, it would turn its hosts into hemophiliacs who would quickly bleed to death and take away the hookworm's meal. A biotechnology company has isolated these molecules and is now trying to turn them into anti-clotting drugs.

• • •

For some parasites, reaching their new home in the body is not enough. Before they can eat and multiply, they build new houses for themselves, using their host's tissue as lumber.

Plasmodium, the parasite that causes malaria, enters the bloodstream through a mosquito bite and lives for a week or so in a liver cell. It then breaks out and gets back into the bloodstream. It rolls and yaws its way in search of its next home, a red blood cell. It's here in the red blood cell that *Plasmodium* can feed on hemoglobin, the molecule that holds on to the oxygen that the red blood cells carry from the lungs. Devouring most of the hemoglobin in a cell, *Plasmodium* can gain enough energy to divide into sixteen new versions of itself, a flock of new parasites bursting out of the cell after two days, all searching the blood for new cells to invade.

Red blood cells are in many ways an awful place to live. Strictly speaking, they're not even cells at all; they're corpuscles. All true cells carry genes in a nucleus and duplicate their DNA in order to become two new cells. Red blood cells originate from cells deep inside our bones. These stem cells, as they're known, divide and take the form of the various components of the blood, such as white blood cells, platelets, and red blood cells. But while other cells get their proper rations of DNA and proteins, red blood cells get no DNA at all. Their job is simple. In the lungs they store oxygen in molecules of hemoglobin. Because oxygen is a powerful atom that can easily react—and damage other molecules—the hemoglobin actually surrounds it by its four chains. Once the red blood cell leaves the lungs and travels through the body, it eventually sets the oxygen free to help the body burn its fuel to produce energy. The cells are simply crates pushed through the circulatory system by a beating heart. If you put white blood cells under a microscope, they reach out lobes to drag themselves across the slide. Red blood cells just sit on the glass.

Because their job is so simple, red blood cells don't need much metabolism. That means they carry few of the necessary proteins for generating energy. Nor do they need to burn fuel and pump out waste. A true cell pumps its fuel in and spits its trash out by means of elaborate channels and bubbles that can shuttle molecules across its outer membrane. A red blood cell has hardly any of this equipment—a couple of channels for water and other essentials—because oxygen and carbon dioxide can diffuse through its membrane without any help. And while other cells have intricate scaffolding inside their membranes to keep them stiff and strong, a red blood cell is the contortionist of the body's cellular circus. It travels three hundred miles in its lifetime, blasted and buffeted by the flow of blood, crashing into vessel walls and getting squeezed through slender capillaries, where it has to travel with other red blood cells in single file, compressed to about a fifth of its normal diameter, bouncing back to its normal size once it's through.

In order to survive the abuse, the red blood cell has a network of proteins undergirding its membrane that are arrayed like the knit of a mesh bag. Each string of proteins making up the mesh is also folded up like a concertina, allowing it to stretch out and squeeze back in response to stress coming from any direction. But as flexible as a red blood cell may be, it can't take this abuse forever. Over time its membrane becomes stiff, and it has a harder time squeezing through the capillaries. It's the spleen's job to keep the body's blood supply young and vibrant. As red blood cells pass through the spleen it inspects them carefully. It can recognize the signs of old age on the surface of red blood cells, like the wrinkles on a face. Only young red blood cells make it out of the spleen; the rest are destroyed.

Despite all of the disadvantages of a red blood cell, *Plasmodium* seeks out this strange empty house. The parasites can't swim, but they can glide along the walls of blood vessels. To do so, they set down hooks on the vessel wall, drag them back to their tail end, and put new hooks down to take their place, like a cellular tank tread. At the parasite's tip are sensors that respond only to young red blood cells, clasping on to proteins on the cells' surface. Once *Plasmodium* fixes on a cell, it latches on and rolls itself over onto its head and prepares to invade.

The head of the parasite is ringed by a set of chambers like the barrel of a revolver. Out of the chambers comes a blitz of molecules in a matter of seconds. Some of the molecules help the parasite push aside the membrane skeleton and work its way inside. The same hooks that acted as the parasite's tank treads while it wandered along the vessel walls now latch on to the edges of the hole and drive the parasite through it. The parasite blasts out sheets of molecules, which join together and form a shroud around the parasite as it goes in. Fifteen seconds after the blast, *Plasmodium*'s back end disappears through the hole, and the resilient meshwork of the red blood cell simply bounces back again, sealing itself shut.

Once inside, the parasite is in the pantry. Each red blood cell's interior is 95 percent hemoglobin. *Plasmodium* has a mouth of sorts on one side—a port that can swing open—and when it does, the outer membrane of the parasite's bubble opens as well, bringing the parasite briefly into contact with the red blood cell's contents. A little dollop of hemoglobin oozes into the maw, which then twists shut. The hemoglobin now floats in a bubble inside the parasite, which contains molecular scalpels that slice apart the molecules. *Plasmodium* makes a succession of cuts that open up their folded branches, letting them fall apart into smaller pieces and capturing the energy that had been held in those bonds. The core of hemoglobin molecules is a strongly charged, iron-rich compound that is poisonous to the parasite. It tends to lodge itself in *Plasmodium*'s membrane, where its charge disrupts the normal flow of other molecules in and out. But *Plasmodium* can neutralize the toxic heart of its meal. It strings some of it in a long, inert molecule called hemozoin. The rest of the compound gets processed by the parasite's enzymes, which reduce its charge and make it unable to penetrate the membrane.

Plasmodium does not live by hemoglobin alone, however. It needs amino acids to build its molecular scalpels, and it also needs them to multiply into sixteen new copies. In those two days, the metabolic rate within an infected cell rises three hundred fifty times, and the parasite needs to make new proteins and get rid of the wastes that it makes as it grows. If *Plasmodium* had infected a true cell, it could simply hijack its host's bio-chemistry for those jobs, but in a red blood cell it has to build the machinery from scratch. In other words, *Plasmodium* has to transform these mere corpuscles into proper cells. Out from its bubble it extends a tangled maze of tubes that reach all the way to the membrane of the red blood cell itself. It's not clear whether *Plasmodium*'s tubes actually punch their way through the membrane of the red blood cell or jack into the channels that are already there. In either case, the parasitized red blood cell can start dragging in the building blocks the parasite needs to grow.

Suddenly crowded with channels and tubes, the surface of the red blood cell starts to lose its springiness. This could be fatal for the parasite, because if the spleen discovers that the cell is no longer its lithe young self, it will destroy it—along with any parasites it may harbor. As soon as it enters the red blood cell, *Plasmodium* releases proteins that are ferried through the tubes to the underside of the cell's membrane. These molecules belong to a common class of proteins found in every sort of organism on Earth. Known as chaperones, they help other proteins fold and unfold properly even when they're being disrupted by heat or acid. In the case of *Plasmodium*'s proteins, though, the chaperones seem to protect the red blood cell from the parasite itself. They help the cell's skeleton stretch out and collapse back tight again, despite the parasitic construction getting in their way.

Within a few hours, the parasite has transformed and stiffened the red blood cell so much that there's no hope in trying to disguise it as a healthy corpuscle. Now the parasite dispatches a new set of proteins to the surface of the cell. Some of them ball up in clumps under the cell's surface, giving the membrane a goose-bumpy look.

Plasmodium then pierces the goose bumps with sticky molecules that can grab hold of receptors on the cells of the blood vessel walls. As these red blood cells stick to the vessel walls they drop out of the body's circulation. Rather than trying to sneak through the slaughterhouse of the spleen, *Plasmodium* evades it altogether. Their red blood cells instead clump up in capillaries in the brain, the liver, and other organs. *Plasmodium* spends another day dividing, until the red blood cell is nothing more than a taut skin around the bulging bundle of parasites. Finally, the new generation of *Plasmodium* breaks out of the cell and looks for new red blood cells to invade. Left behind in the dead cell is a clump of used-up hemoglobin. For a time the cell was the parasite's home, a cell like none other in the human body, but in the end it becomes its garbage dump.

•   •   •

Trichinella is also a biological renovator, and in some ways it's more impressive than *Plasmodium:* it's a multicellular animal that can live inside a single cell. When this worm hatches from an egg in its host's gut, it drills through the intestinal wall and travels the body through the circulatory system. It follows the flow into the fine capillaries, where it leaves the bloodstream and works its way into the muscles. It crawls along the long muscle fibers and then penetrates one of the long, spindle-shaped cells that make them up. In the 1840s, when scientists first recognized *Trichinella*'s cysts lodged in muscles, they thought the tissue had degenerated and that the parasite slept inside, simply waiting to reach its final host. At first, the invaded muscle cell does seem to atrophy. The proteins that serve as the scaffolding of the cell and make it rigid fade away. The muscle's own DNA loses its power to make new proteins, and within a few days after the worm has entered, the muscle changes from wiry to smooth and disorganized.

But the parasite is only tearing down the cell so that it can rebuild it. *Trichinella* doesn't disable its host's genes—in fact, they start copying themselves until they've quadrupled. But this abundance of genes now follows *Trichinella*'s commands, making proteins that will turn the cell into a proper home for the parasite. Scientists once thought this kind of genetic control was limited to viruses, which use their host's DNA to make more copies of themselves. *Trichinella*, they now realize, is a viral animal.

Trichinella turns the muscle cell into a parasite placenta. By making the muscle cell loose and flexible, the parasite makes room on its surface for new receptors for taking in food. The parasite also forces the cell's DNA to churn out collagen, which forms a tough capsule around the cell. It makes the cell produce a signal molecule known as vascular endothelial growth factor. This molecule normally sends a signal to blood vessels to grow new branches in order to help heal wounds or nurture growing tissues. *Trichinella* uses the signal for its own purposes: to weave a mesh of capillaries around it, using the collagen capsule as their mold. Through the vessels comes a nourishing flow of blood, allowing the parasite to grow and swell inside its muscle cell, which bulges and groans as the worm rocks back and forth and probes its little home.

Parasites can also reconstruct the interiors of plants as drastically as they can those of animals. It may come as a surprise that plants actually have parasites at all, but they're positively overrun with them. Bacteria and viruses live happily in plants, sharing them with animals, fungi, and protozoa. (Trypanosomatids, close relatives of the parasites that give us sleeping sickness, can live inside palm trees.) Plants are even hosts to parasitic plants that drive their roots into their hosts. Parasitic plants come into this life lacking at least some of the skills that a plant needs to live on its own. Bird's beak, which lives in salt marshes, is a part-time parasite that has to steal fresh water from pickleweed and other plants that can get rid of the salt; they can handle their own photosynthesis and get their own soil nutrients. Mistletoe can photosynthesize, but it can't draw its own water and minerals from the soil. Broomrape can do nothing for itself.

There are also millions of species of insects and other animals that live on plants, but before 1980, few ecologists thought of them as parasites. They were considered herbivores, essentially little spineless goats. But Peter Price, an ecologist at Northern Arizona University, pointed out that there's a fundamental difference between these animals and herbivores. Herbivores are to plants as predators are to prey: an animal that can eat any number of species. A coyote will be happy with a bat, a rabbit, or a cat, while a sheep is equally easy about the plants it eats, entering a field and devouring the clover, the timothy, the Queen Anne's lace. Some insects, like woolly bear caterpillars, graze like sheep, taking small bites from individual plants of different species and moving on. But many insects are limited to only one plant, at least for one stage of their life. A caterpillar that goes from egg to pupa on a single milkweed plant is no different from a tapeworm, which can live as an adult only in the intestines of a human. And many plant-eating insects spend their entire lives on a single plant, shaping their lives to that of their host.

One of the most powerful demonstrations of Price's argument is nematodes that live in plant roots. These parasites are spectacular pests, destroying 12 percent of all the cash crops in the world. One particular kind—root-knot nematodes of the genus *Meloidogyne*—are also an uncanny botanical reflection of *Trichinella*. Each nematode hatches from an egg in the soil and crawls to the tip of a root. It carries a hollow spike in its mouth, which it stabs into the root. Its saliva makes the outer cells burst, freeing up a space through which the nematode can slip. It nudges its way between the cells inside the root until it reaches the root's core.

The nematode then pierces a few cells around it, injecting a peculiar poison into them. The cells start making copies of their DNA, and the extra gene starts making a flurry of proteins. Genes switch on in these root cells that would never normally become active. The job of a root cell is to pull in water and nutrients from the soil and pump them into a plant's circulatory system, a network of tubes and cavities that carries the food to the rest of the plant. But under the spell of the nematode, a root cell starts working backward. It begins to suck in food from the plant. Its cell walls become leaky enough to let the food flow in easily, and it sprouts fingery ingrowths, where it can store the food. The nematode spits molecules into the altered cell, which form themselves into a sort of intercellular straw, which it uses to suck up the food being pumped in from the rest of the plant. As the cell swells with food, it threatens to burst the entire root open. To protect it, the nematode makes the surrounding cells multiply and form a sturdy root knot to withstand the pressure. Just as *Trichinella* speaks the genetic language of mammals, root nematodes have learned the language of plants.

•   •   •

Parasites live in a warped version of the outer world, a place with its own rules of navigation, of finding food and making a home. While a badger digs itself a den or a bird weaves itself a nest, parasites often act as architects, casting a biochemical spell to make flesh and blood change into the form they desire, a heap of planks swirling together into a house. And inside their hosts, parasites also have their own bizarre inner ecology.

Ecologists study how the millions of species on Earth share the world, but rather than take on the whole planet at once, they generally focus on a single ecosystem, be it a prairie, a tidal flat, or a sand dune. Even within those limits, they are frustrated by loose frontiers, by the way seeds blow in from miles away or wolves lope in from the other side of a mountain. As a result, ecologists have done some of their most important work on islands, which may be colonized only a few times over the course of millions of years. Islands are nature's own isolated laboratories. On them, ecologists have figured out how the size of a given habitat determines how many species can survive on it. And they've taken that knowledge back to the mainland, showing how a fragmented ecosystem becomes its own archipelago, where extinctions can strike.

To a parasite, a host is a living island. Bigger hosts tend to have more species of parasites in them than small ones, just as Madagascar has more species than the Seychelles. But as islands go, hosts have some quirks. Parasites can find in them a vast number of ecological niches, because a body has so many different places to which they can adapt. On the gills of a single fish, a hundred different species of parasites may each find their own niche. An intestine may look like a simple cylinder, but to a parasite, each stretch has a unique combination of acidity, of oxygen levels, of food. A parasite may be designed for living on the surface of the intestines, inside the film that coats it, or deep among its fingerlike projections. In the bowels of a duck, fourteen species of parasitic worms may live (their combined population is on average twenty-two thousand), and each species takes as its home a particular stretch of intestine, sometimes overlapping with its neighbors, often not. Parasites can even find a way to parcel out the human eye: one species of worm in the retina, one in the chamber, one in the white of the eye, one in the orbit.

In hosts where parasites can find enough niches, they don't compete over their island of flesh. But when they all want the same niche, ugliness usually breaks out. A dozen species of flukes may be able to infect a single snail, for example, but they all need to live in its digestive gland to survive. When parasitologists crack open the shells of snails, they typically don't find those dozen species of flukes inside, but several individuals from one species. The flukes may devour their competition or release chemicals that make it harder for newcomers to invade. Other parasites living inside other animals can also compete with one another. When thorny-headed worms arrive in a rat's intestines, they drive tapeworms out of the most fertile region, exiling them down into a stretch of the bowels where it's much harder to find food.

The most vicious and unneighborly behavior of all, though, can be found among some of the parasitic wasps that so impressed Darwin. This shouldn't come as too much of a surprise, given the gruesome way the wasps treat their hosts. The mother wasp roams over the countryside, sniffing the air for the scent of the plants its host—often a caterpillar but sometimes another insect such as an aphid or an ant—feeds on. Once it gets closer, it sniffs for the scent of the caterpillar itself or its droppings. Parasitic wasps alight on their host and jam their stinger into the soft section between the plates of the caterpillar's exoskeleton. Their stinger isn't actually a stinger at all, though; it is actually called an ovipositor, and it delivers eggs—in some cases just a handful, in others hundreds. Some wasps also inject venom that paralyzes their hosts, while others let them go back to feeding on leaves and stems. In either case, the wasp eggs hatch, and larvae emerge into the caterpillar's body cavity. Some species only drink the caterpillar's blood; others also dine on its flesh. The wasps keep their host alive for as long as they need to develop, sparing the vital organs. After a few days or weeks, the wasp larvae emerge from the caterpillar, plugging up their exit holes behind them and weaving themselves cocoons that stud the dying host. They mature into adult wasps and fly away, and only then does the caterpillar give up the entomological ghost.

When different species of wasps compete for the same caterpillar, it can become a brutal struggle. A clutch of wasp larvae may end up stunted and starved if they face too much competition, and the danger is worse for wasps that need a long time to mature in caterpillars. The wasp *Copidosoma floridanum* takes an entire month to mature inside the cabbage looper moth. As a result, it is a staggeringly unfriendly parasite.

Typically, *Copidosoma* lays only two eggs in its host, one male and one female. As with any egg, each begins as a single cell and divides, but then it veers away from the normal path of development most animals follow. The cluster of wasp cells divides itself up into hundreds of smaller clusters, each of which then develops into separate wasps. Suddenly, a single egg gives rise to twelve hundred clones. Some of the clusters develop much faster than the rest, becoming fully formed larvae only four days after their original egg was laid. These two hundred larvae, known as soldiers, are long and slender females, with tapered tails and sharp mandibles. They roam through the caterpillar, seeking out one of the tubes the caterpillar uses to breathe. They wrap their tails around a breathing tube, and like sea horses anchored to a coral reef, they rock in the flow of caterpillar blood.

The task for these soldiers is simple: they live only to kill other wasps. Any wasp larva that passes by, whether other *Copidosoma floridanum* or another species, prompts a soldier to lash out from its tube, snagging the larva in its mandibles, sucking out its guts, and letting the emptied corpse float away. As this slaughter goes on, the rest of the *Copidosoma* embryos slowly develop and finally grow into a thousand more wasp larvae. These larvae, called reproductives, look very different from the soldiers. They have only a siphon for a mouth, and they're so tubby and sluggish that they can move only by being carried by the flow of the caterpillar's blood. Reproductives would be helpless against any attack, but thanks to the soldiers, they can just drink the caterpillar's juices as the shriveled corpses of their rivals float past.

After a while, the soldiers turn on their siblings—more specifically, on their brothers. A mother *Copidosoma* lays one male egg and one female egg; after they've both multiplied, they produce a fifty-fifty split betwen the sexes. But the soldiers selectively kill the males so that the vast majority of survivors are females. Entomologists once documented two thousand sisters and a single brother *Copidosoma* emerging from a caterpillar.

The soldiers turn on their own brothers for sensible evolutionary reasons. Males do nothing for their future offspring beyond providing sperm. *Copidosoma's* hosts are hard to find—they are spread out like islands separated by miles of ocean, so males that emerge from a caterpillar will probably mate successfully close to home with their sisters. In such a situation, only a few males are necessary, and any more would mean fewer females for them to mate with, and fewer offspring. By killing the male reproductives, the female soldiers ensure that the host will be able to support the most females possible and help carry on the genes they share with their sisters.

As ruthless as soldiers may be, they're also selfless. They are born without the equipment for escaping the caterpillar themselves. While their reproductive siblings drill out of the host and build themselves cocoons, the soldiers are trapped inside. When their host dies, they die with it.

Making that final journey—leaving the host—is the most important step in a parasite's existence. It takes particular care to be ready to get out when the time is right, because otherwise it will be doomed to die with its host. That's why people who need to be tested for elephantiasis, as Michael Sukhdeo was as a child, have to be tested at night. The adult filarial worms live in the lymph channels, and the baby worms they produce move into the bloodstream, spending most of their time in the capillaries in tissues deep within the body. But the only way for a baby worm to grow to adulthood is to be taken up in the bite of mosquitoes that come out at night. Somehow, deep inside our bodies, the worms can figure out what time of the day it is—perhaps by sensing the rise and fall of their host's body temperature—and move out into the blood vessels just under the skin, where they're likely to get sucked up by a mosquito. By two in the morning, the worms that haven't been picked up in a bite start moving back to their host's core to wait for the next dusk.

Parasites can also use hormones to signal them when it's time to leave. The fleas on a female rabbit's skin can detect hormones in the blood they drink from her. They can tell when she's about to give birth, and they respond by rushing to the front of her face. Once she has delivered her babies and is nuzzling and licking them, the fleas leap onto the newborns. Baby rabbits can't groom themselves yet, and their mothers clean them only when they visit their nest once a day to nurse. That makes the baby rabbits wonderfully tranquil homes for fleas. The fleas immediately start feeding on the babies, mating, and laying eggs. The new generation of fleas grows up on the babies, but when they sense that the mother is pregnant again, they hop back on her. There they wait to infect her next litter.

Getting to a new host can become a huge challenge when a parasite's species of choice is a solitary creature. Dig a few feet down into the hard summer dirt of an Arizona desert, for example, and you may a find a toad. It is the spadefoot toad *Scaphiopus couchi*, and it is sleeping away the eleven-month drought that dominates every year. It sits underground, not eating, not drinking. Its heart barely beats, but its cells still have to purr metabolically along, and it stores its wastes in its liver and bladder. In July or August the first rains come, monsoons that roar down and break up the soil. On the first wet night the toads come alive and crawl out.

The toads gather in ponds, where the males outnumber the females ten to one. They attract the females by singing in floating choruses, croaking so passionately that their throats bleed. A female drifts among the males until she finds the voice she likes and nudges the male. He climbs on her and they lock together, the female letting slide a raft of eggs that the male fertilizes with his sperm. By four in the morning the courtship is over. Before the hot sun rises, the toads have crawled back down a few inches into the ground. Only when the sun sets again (and only if there's enough water) will the toads return to the surface. When they aren't mating, the toads are eating enough food to tide them over for the rest of the year. A toad can eat half its weight in termites in one night. Meanwhile, their offspring grow frantically from egg to toadlet in only ten days, since the rainy season is only a few weeks long. As the rains taper off the toads all disappear underground, having spent a few days out of the earth, and return to their life of sleep.

With so little opportunity to go from host to host, a spade-foot toad might seem a bad choice for a parasite. There are, in fact, hardly any parasites that have gotten a foothold inside the spadefoot, and most of them can only mount feeble infections. But one parasite positively revels in the spadefoot life, a worm named *Pseudodiplorchis americanus*. *Pseudodiplorchis* belongs to a group of parasites called monogeneans, delicate blobby worms that almost always live on the skin of fish and travel from host to host in the comfort of ever-present water. Yet, half of spade-foot toads carry the monogenean *Pseudodiplorchis*, and each toad carries an average of five.

Of all places, *Pseudodiplorchis* chooses the toad's bladder to live during the long sleep. As the toad pumps more salts and other wastes into the bladder the parasite goes on with its life, sucking blood and mating. Within each female *Pseudodiplorchis*, hundreds of eggs mature into larvae. They sit inside her for months, waiting for the toad to rouse. The parasites will wait as long as the toad waits, even if the rains don't come until the next year. When the rains do fall, the parasite is caught in a del-uge of its own. After the toad has clawed its way to the ground, its skin soaks up water, which floods through its bloodstream, scouring out all the poisonous waste that has built up in its body over the year, through its kidneys and into its bladder. This tor-rent of urine suddenly turns the parasite's habitat from a salty ocean to a freshwater pool. *Pseudodiplorchis* holds tight during the torrent and goes on waiting. It waits out the male choruses and the female inspections. Only when their toad host is sexu-ally aroused as it tries to mate with another toad does a mother *Pseudodiplorchis* send her hundreds of young rushing out of the bladder and into the pond. When they reach the water, they rip out of their egg sacs and swim free.

Now, after their eleven-month wait, the parasites have to

race. They have only a few hours to find another host in the mating pool before the toads crawl back underground and the sun rises and any stranded parasites fry. As they swim through the pond they have to be sure that they don't crawl onto one of the other species of desert toads that crowd the water as well. Some kind of unique skin secretion from the spadefoot probably guides them to their host. *Pseudodiplorchis* has an awesome homing ability in its ponds. For many parasites, it's not unusual for only a few out of thousands of larvae to find a host in which they can mature. *Pseudodiplorchis* has a success rate of 30 percent. As soon as it hits its host, a *Pseudodiplorchis* larva starts crawling up the toad's side. It comes out of the water altogether, climbing as high as it can go. It ends up on the toad's head, and once there, it can find the nostrils and slip inside.

The race goes on further: *Pseudodiplorchis* still has to get into the toad's bladder before the rainy season ends. And within the toad, *Pseudodiplorchis* faces conditions just as murderous as the desert sun. It travels down the toad's windpipe, drinking blood as it goes, until it gets to the lungs. There it lives for two weeks, fighting off the toad's efforts to cough it up, maturing into a young adult about a tenth of an inch long. It leaves the lungs and crawls into the toad's mouth, only to turn around and dive down its esophagus and into its gut.

The acids and enzymes the toad uses to digest its food should dissolve such a delicate parasite. If you pull a newly arrived *Pseudodiplorchis* out of a toad's lung and stick it directly into its intestines, the parasite will die in minutes. But in its two weeks in the lungs, it can prepare itself for the trip by storing up a collection of liquid-filled bubbles in its skin. When it dives into the toad's digestive tract, it lets the bubbles burst, spilling out chemicals that neutralize the compounds trying to digest it. Yet, even with this protection, *Pseudodiplorchis* doesn't dawdle: it charges through the entire digestive tract of the toad in half an hour and makes its way into the bladder. The entire trip, from nose to lung to mouth to bladder, takes no more than three weeks, and by then the host toad has finished its annual mating and feasting and is back underground.

The spadefoot toad is one of the few hosts that leads a life as isolated as its parasites; together they spend a year in the ground waiting for the chance to see their kind again.

• • •

Parasites have colonized the most hostile habitats nature has to offer, evolving beautifully intricate adaptations in the process. In this respect, they're no different from their free-living counterparts, much as that might horrify Lankester. And I haven't even had room in this chapter to talk about the most remarkable adaptation that parasites have made: fighting off the attack of the immune system. That fight demands a chapter of its own.

3

The Thirty Years' War

O Rose, thou art sick.
The invisible worm
That flies in the night,
In the howling storm,

Has found out thy bed
Of crimson joy,
And his dark secret love
Does thy life destroy.

—William Blake, "The Sick Rose"

A man came one day to the Royal Perth Hospital in Australia saying he was tired. He had been tired for two years, and now, in the summer of 1980, he decided it was time to find out what was wrong with him. His health wasn't perfect, but it wasn't terrible either. He had been a heavy smoker in his teens and twenties, but at forty-four, his only indulgence was a glass of white wine each night.

His doctor could feel through his skin that his liver was swollen. On an ultrasound image, two of its three lobes loomed too large. Yet, there were no signs of the kinds of trouble the doctor would expect to find, such as a tumor or cirrhosis. It was when the doctor got the report on the man's stool that he realized what had happened: the stool was loaded with the spiny eggs of *Schistosoma mansoni*—blood flukes found only in Africa and Latin America.

The doctor had the man walk him through his life. It had started roughly. He had been born in Poland in 1936. The Soviet army had taken his family during the Second World War and held them in a Siberian prison camp. Toward the end of the war they had escaped, traveling through Afghanistan and Persia, finally ending up in a refugee camp in East Africa. For six years, savannas were his playgrounds, until 1950, when his family emigrated to Australia. He had remained there for the rest of his life.

The math is simple enough, yet hard to believe: the only time in the man's life when he was anywhere near *Schistosoma mansoni* was in the late 1940s. When he swam and bathed in Tanzanian lakes, at least one pair of flukes had invaded his skin and journeyed into his veins; they had traveled with him to Australia and started a new life with him, and male and female flukes had gone on living, quietly entwined and pumping out eggs, for over thirty years.

What makes the longevity of the blood flukes all the more impressive is that they attained it under perpetual menace and attack. Lankester was under the impression that once inside a host, a parasite was home free. It needed to do nothing more than drink up the food that bathed it, and could in fact do nothing more. But he wrote his essay "Degeneration" in 1879, when immunology, the science of the body's defenses, was still little better than alchemy. Physicians knew that they could protect people from smallpox by injecting a bit of a pox sore into them, but they had no idea how they were actually saving lives.

The Thirty Years' War

Within a few years of Lankester's essay, scientists would discover predatory cells roving our bodies and devouring bacteria, and immunology was born.

To sum up what scientists have learned since then about the immune system is like trying to reproduce the Sistine Chapel in crayon. It is orchestral in its complexity, with a huge diversity of cells, all communicating among each other with a dictionary's worth of signals, along with dozens of kinds of molecules designed to help the cells decide what should be destroyed and what should be spared. It acts like a blood-borne brain. But here, at any rate, is a brief survey of the most important ways in which our bodies kill parasites.

The immune system attacks an intruder—bacteria crawling into a cut, for instance—in a succession of waves. One of the first waves is a collection of molecules called complement. When complement molecules hit the surface of bacteria, they latch on and change their shape so that they can snag other passing complement molecules. Gradually the molecules build up on the surface. They assemble themselves into tools of destruction, like drills that can open a hole in the bacteria's membranes. They also act like beacons, making the bacteria more visible to immune cells. Complement molecules also land on our own cells, but they do no harm. Our cells are coated with molecules that can clamp onto a complement molecule and cut it apart.

Also arriving early at the cut are wandering immune cells, the most important of which are the macrophages. They have some crude ways of recognizing bacteria if they happen to bump into them, and they can suck the invaders into their cores and slowly digest them. At the same time, the macrophages also release signals that bring the rest of the immune system's attention to the site. Some of these signals make the infection swell up by loosening the neighboring blood vessel walls. That lets other immune cells and molecules flood into the tissue. The signaling molecules released by the macrophages also latch onto immune cells that happen to be flowing by in nearby blood vessels. They lead the cells through the vessel wall and to the infection, like a boy dragging his mother by the hand down a toy store aisle.

With enough time, the immune system can organize a new wave of attack, using much more sophisticated cells: B and T cells. Most of our cells come with a standard issue of receptors on their surface. One red blood cell looks pretty much like the next. But when B and T cells form, they shuffle the genes that make the receptors on their surface. The cells use the altered genes to build new receptors with shapes not found in any other immune cell. This shuffling can produce hundreds of billions of different shapes, so that each new B or T cell is as distinct as a human face.

Because they are so diverse, B and T cells can grab a huge range of molecules, including the ones on the surface of invaders. (Foreign molecules that trigger an immune response are called antigens.) First, though, the cells have to get a proper introduction to the antigens. This job is accomplished by macrophages and other immune cells. As they engulf bacteria or their cast-off fragments the immune cells cut them up into little pieces. They then bring these antigens to their surface, displaying them in a special cup (the major histocompatibility complex, or MHC for short). Parading these conquests, the immune cells travel into the lymph nodes. There they bump into T cells. If a T cell has the right kind of receptor, it can lock onto the antigens displayed by a macrophage. As soon as it recognizes the antigen, the T cells start multiplying quickly into a battalion of identical cells, all equipped with the same receptor.

These T cells can take one of three forms, each of which kills the invaders in a different style. Sometimes they become killer T cells, which search the body for cells that have been invaded by pathogens. They recognize infected cells, thanks again to MHC. Like macrophages, most cells in the human body can display antigens on MHC receptors of their own. If the killer T cell recognizes these signs of trouble, it commands an infected cell to commit suicide. The parasite within dies along with it.

In other cases, activated T cells set out to coordinate other immune cells to do a better job of killing. Sometimes they help by becoming inflammatory T cells. These cells crawl their way to the macrophages that are struggling to fight the rising tide of bacteria. They lock onto the antigen displayed on the macrophage's MHC. That locking acts like a trigger, turning the macrophage into a more violent killer, spraying more poisons. At the same time, the inflammatory T cells help make the cut swell far more than the macrophages can manage on their own. The inflammatory T cells also kill off tired old macrophages and spur the production of new ones to devour their elder cousins. They're like battle-hungry generals: they're good to have around in a war but can't be allowed to get out of control. Too much inflammation, too many poisons created by macrophages, and the immune system will start destroying the body itself.

In the third form that T cells take, they help B cells make antibodies. B cells have the same diversity of surface molecules as T cells, so they also have the potential to snag onto billions of different kinds of antigens. After a B cell has latched onto a fragment, a helper T cell may come along and hook onto it at the same time. In these unions, the T cell can give the B cell signals to start making antibodies. Antibodies are a kind of free-floating version of a B cell receptor, also able to clasp onto an antigen from an invader.

Once they're activated, B cells spew antibodies out into the body, and depending on the particular antibody, they can fight the infection in several ways. They can cluster around a toxin made by bacteria and neutralize it. They can help the complement molecules trying to drill into the bacteria to make bigger holes. They can latch onto bacteria and foul up the chemistry they use to invade the body's cells. They can tag bacteria to make them a clearer target for macrophages.

As the majority of B and T cells go about eradicating the bacteria from the cut, a few sit out the attack. These are known as memory cells; it is their job to preserve a record of the invader for many years after the infection. If the same kind of bacteria should get into the body again, the memory cells can switch back on and orchestrate a swift, overwhelming assault. These cells are the secret to vaccines. Even if immune cells are exposed only to an antigen, they can produce memory cells. Because a vaccine contains only a molecule and not a living organism, it doesn't make a person sick, but it can still prime the immune system to wipe out the pathogen if it ever meets up with it again.

T cells, B cells, macrophages, complement molecules, antibodies, and all the other parts of the immune system form a tight net that perpetually sweeps our bodies clean. Every now and then, though, a parasite slips through and establishes itself. Its success isn't simply due to some oversight but to the parasite's ability to escape the immune system. Bacteria and viruses have their own tricks, but many of the most intriguing strategies are found among the "classic" parasites—the protozoa, flukes, tapeworms, and other eukaryotes. They can evade the immune system, distract it, wear it out, and even take control of it, confusing its signals into a weakened state or, if need be, a heightened one. One sign of the sophistication of these parasites is the fact that there is still no vaccine for them, while there are many vaccines for viruses and bacteria. If Lankester had known any of this, perhaps he wouldn't have given parasites the bad reputation they still haven't been able to shake.

• • •

In September 1909, a strong young man from Northumberland came down with sleeping sickness in northeastern Rhodesia, near the Luangwa River. His illness wasn't diagnosed for two months, but soon afterward he arrived back in England, and was treated by doctors at the Liverpool School of Tropical Medicine. He was admitted to the Royal Southern Hospital on December 4, where his doctor was Major Ronald Ross. Ross was one of the giants of tropical medicine, who a decade earlier had figured out the cycle of malaria: the way *Plasmodium* travels between mosquito and human. The sleeping sickness patient's blood was seething with the trypanosome parasites, thousands of augur-shaped creatures to every drop. His glands swelled, and his legs became covered in rashes. For weeks he dwindled. Ross tried to destroy the parasites with an arsenic compound but had to stop when it damaged the man's eyes instead. In April, the patient vomited for four days and lost ten pounds. From then on, he became drowsier and drowsier, although he would occasionally perk up. His liver expanded, and the vessels in his brain became congested.

Ross began trying out other treatments. He inoculated a rat with the blood from his patient, let the parasites multiply, and then drew off some of the rat's blood. He heated it to kill the trypanosomes, and then injected this crude vaccine back into the man. It did nothing. In May, his patient's anal sphincter became paralyzed and Ross was sure he was going to die, but a week later he went through a sudden remarkable improvement. It lasted only a few days before he faded again, came down with pneumonia, and passed away. At the autopsy, Ross couldn't find a single trypanosome.

A few years earlier, Ross had invented a quick way to detect blood parasites, and he used the method on the patient during his final three months. In the process he got the world's first day-by-day portrait of sleeping sickness. He plotted it out on what he described in a report on his patient as a "remarkable graph." The graph showed a clear rhythm: for a few days the trypanosomes would skyrocket, multiplying by as much as fifteen-fold. Then, just as suddenly, they would drop back down to barely detectable numbers. The cycle would take a week or so, and the man's fevers and changing white blood cell counts followed in its wake. The man hadn't been attacked by a single assault of parasites—a string of outbreaks had flared and died within him.

Ross saw in his patient "a struggle between the defensive powers of the infected body and the aggressive powers of the trypanosomes." Exactly what the nature of that struggle was he couldn't say. With another ninety years of study, scientists still can't make a sleeping sickness vaccine, but they do at least understand how trypanosomes ride their spiky wave until their host dies. They play an exhausting game of bait-and-switch.

If you could fly *Fantastic Voyage*-style over a trypanosome, you'd be bored with the view. It would be like the drabbest cornfield in Iowa: millions of stalks all crammed together with barely a space between them. Fly to the next trypanosome and there's no relief: the cornstalks would be identical with the first one. In fact, go to any of the millions upon millions of trypanosomes in a human host at any given moment, and you'll most likely find the same coat.

For a human immune system, these parasites should be as easy to kill as fish in a barrel. If the immune system learns how to recognize only one of these cornstalk molecules, it can attack just about every parasite in the body. And indeed, as a host's B cells begin to produce antibodies tailored to the cornstalks, the trypanosomes start to die. But not completely. Just when it looks as though the trypanosomes are about to disappear into obscurity, their numbers bottom out and rise again. The view has changed. If you were now to fly over the trypanosomes, you'd find not corn but wheat—an utterly drab expanse, but a completely different kind of expanse.

The quick change happens thanks to the unique way the trypanosome's genes are laid out. The instructions for building the molecule that makes up the trypanosome's coat sit on a single gene. Normally, when the trypanosome divides, the new parasites use that same gene to make the same coat. But once every ten thousand divisions or so, a trypanosome will abruptly retire the gene, cutting it out of its position in the parasite's DNA. It then reaches into a reserve of a thousand other coat-building genes, selects one, and pastes it into the old gene's position. The new gene starts making its surface molecule: a molecule that's similar to the previous one, but not identical with it.

Now the immune system, so focused on the first coat, needs time to recognize the second one and make new antibodies for it. In that time, the trypanosomes with the new coat are safe, and they can multiply furiously. By the time the immune system catches up and is attacking the trypanosomes with a new antibody, another trypanosome has installed a third gene and is making a third coat. The chase goes on for months or years, the trypanosomes flinging off their coats and putting on new ones hundreds of times. With so many different kinds of trypanosome fragments building up in the bloodstream, the host's immune system becomes chronically overstimulated, attacking its own body until the victim dies.

This bait-and-switch strategy works only because the parasite can dip into a reservoir of coat-producing genes. But these genes can't be called from their bullpen in any random order. Say that the first generation of trypanosomes to get into a person's body were to switch on all their coat-building genes. The immune system would make antibodies to all of them and bring the infection to a quick stop. And if a new generation of parasites were to turn back to an old coat gene, the immune system would still have some antibodies left over with which it could fight it. Instead, the trypanosomes carefully go through their lineup in a predetermined order. Take two trypanosome clones and infect two mice with them, and their descendants will switch on the same genes in the same order. That way, the parasite can stretch out its infection for months.

Ronald Ross is remembered today for his work on malaria rather than sleeping sickness. Yet he never managed to discover much about the way *Plasmodium* fights the human immune system. Trypanosomes flaunt their evasions through their booms and busts, but *Plasmodium* is subtler. For much of its time in the body, the parasite runs from one cover to the next. When it firsts enters a human through a mosquito bite, it can get to the liver in half an hour, which is often fast enough to escape the notice of the immune system. The parasite slides into a liver cell to mature, and here it comes to the body's attention. The liver cells grab stray proteins from *Plasmodium* floating inside them, cut them up, and shuttle them up to their surfaces, where they display them on their MHC molecules. The host's immune system recognizes these antigens and starts organizing an attack against the sick liver cells. But the attack takes time— enough time for the parasite to divide into forty thousand copies in a week, burst out of the liver, and seek out blood cells. By the time the immune system is ready to destroy infected liver cells, the cells have become empty husks.

Meanwhile, the parasites are invading red blood cells and making their home improvements. *Plasmodium* has to go to a lot of effort to make up for the cells' lack of genes and proteins, but their barrenness has its advantages as well: a red blood cell is a good place to hide. Because they don't have genes, they can't make any MHC molecules, so they have no way of showing the immune system what's inside them. For a time, *Plasmodium* can enjoy perfect camouflage inside the cell.

As the parasite divides and fills the cell it has to start supporting the membrane with its own proteins. To avoid being destroyed in the spleen, it builds knobs on the surface of the cell, each with little latches that can snag onto the walls of blood vessels. These latches pose a danger of their own: they risk getting the attention of the immune system. Antibodies can be made against them, and an army of killer T cells can be assembled that recognizes these signs of an infected cell.

Because these latches can be recognized by the immune system, scientists have spent a lot of time studying them in the hope of building a vaccine against malaria. In the 1990s they were able for the first time to sequence the genes that carry the instructions for the latches. They found that it takes only a single gene to make a latch, but there are over a hundred different genes in *Plasmodium* DNA that can make one. And while every sort of latch can hook the red blood cell to a blood vessel wall, each one has a unique shape.

When *Plasmodium* first invades a red blood cell, it switches on many of these latch-making genes at once, but the parasite selects only one kind of latch to put on its surface. The red blood cell thus will be covered with that particular style of latch alone. When the cell ruptures, sixteen new parasites emerge and they will almost always use the same gene to make the same latch. But every now and then, a parasite will switch to another gene and make new latches that are unrecognizable to the immune system. And that's how *Plasmodium* manages to hide in plain sight: by the time the immune system has recognized its latches, the parasite is making new ones. In other words, malaria uses a bait-and-switch strategy very much like the one used by sleeping sickness. Although Ronald Ross didn't know it, his patients struggling against sleeping sickness and malaria were losing to the same exhausting game.

Plasmodium is only one of many parasites that live inside our cells. Some can live in any kind of cell, while others choose only one. Some even specialize in the most dangerous cells of all, the macrophages whose job it is to kill and devour parasites. In this last category is the protozoan *Leishmania*. There are a dozen species of this parasite all of which are carried from person to person by biting insects called sand flies. Each species causes a disease of its own. *Leishmania major* causes Oriental sore—an annoying blister that heals itself like a canker. *Leishmania donovani* attacks the macrophages inside the body and can kill its host within a year. And a third *Leishmania* parasite, *Leishmania brasiliensis*, causes espundia, in which the parasite chews away at the soft tissue of the head until its victim is faceless.

Leishmania doesn't have to muscle its way into its host macrophage the way *Plasmodium* pushes into red blood cells. It's more like an enemy spy that knocks at the door of police headquarters and asks to be arrested. When the parasite is injected during a sand fly's bite, it attracts complement molecules that try to drill into its membrane and attract macrophages to devour it. *Leishmania* can stop complement from drilling into it, but it doesn't destroy the molecule. It still lets complement do its other job: to act like a beacon. A macrophage crawls over the parasite, detects the complement, and opens a hole in its membrane to engulf *Leishmania*.

The macrophage swallows up the parasite in a bubble that sinks into its interior. Normally, this would become a death chamber for a parasite. The macrophage would fuse that bubble with another one filled with molecular scalpels, which it would use to dismantle *Leishmania*. But somehow—scientists still don't know how—*Leishmania* stops the bubbles from fusing. Its own bubble, now safe from attack, becomes a comfortable home where the parasite can thrive.

Leishmania not only alters the particular macrophage it's inside but changes the body's entire immune system. When young T cells encounter antigens for the first time and lock onto them, they can become helper T cells. Which type of helper they become—the inflammatory kind or the kind that helps B cells make antibodies—depends on the balance of certain signals floating through the body. At first, both kinds of T cells start to multiply, but as they do they interfere with one another. In many infections, this struggle tips the balance in favor of one kind of T cell or the other. The winning side launches its own kind of war against the parasite.

Leishmania has figured out how to fix this fight. Clearly, the best way to destroy this parasite would be to make lots of inflammatory T cells. These cells could help the macrophage kill parasites they have swallowed. And that seems, in fact, to be what happens inside people who manage to fight off *Leishmania*. Parasitologists have run experiments in which they infected mice with *Leishmania* and siphoned off the inflammatory T cells made by the mice who survived the disease. The parasitologists then injected these T cells into mice that had been genetically stripped of most of their immune system. The injection let the helpless mice fight off the parasite as well.

But often our bodies can't raise the right defense, and that failure seems to be *Leishmania*'s doing. Sitting inside its host macrophage, it forces the cell to release the signals that tip the immune system in favor of the T cells that help make antibodies. Since *Leishmania* is safely hidden inside macrophages, the antibodies can't reach them. And so the disease goes unchecked.

Plasmodium and *Leishmania* are fussy about where they live, able to survive only in certain types of cells. Most parasitic protozoa are equally choosy, but there are a few that can invade just about anything. One such species is *Toxoplasma gondii*, a creature that lives in undeserved obscurity. Few people know about *Toxoplasma*, even though there's a fair chance that they are carrying it by the thousands in their brains. A third of all the people in the world are infected by it; in parts of Europe almost everyone is a host.

Although billions of humans carry *Toxoplasma*, we are not actually the parasite's natural host. Normally it cycles between cats, domestic and wild, and the animals they eat. The cat releases *Toxoplasma*'s egg-like oocysts in its feces, and the oocysts can wait in the ground for many years to be picked up by an animal such as a bird, a rat, or a gazelle. In their new host, the oocysts hatch and the protozoa move through the body and look for a cell to make their home.

Toxoplasma is a close relative of *Plasmodium*, the protozoan that causes malaria, and it also is equipped with the same special machinery around its tip that blasts its way into a cell. But while *Plasmodium* can live only in liver cells and then red blood cells, *Toxoplasma* doesn't much care. It muscles its way into just about any type of cell.

Once *Toxoplasma* has invaded a cell, it starts feeding and reproducing. After it has divided into 128 new copies, it tears the cell open, and the new parasites spill out, ready to invade fresh cells. After a few days, the parasite shifts gears. Now, instead of invading cells, it builds shells, each of which hides a few hundred *Toxoplasma* individuals. Every now and then, one of the cysts will break open and the parasites inside will invade cells and produce new *Toxoplasma*. But their descendants promptly build cysts of their own and vanish into them. There they will sit for years, until their host is eaten by a cat. Once inside their final host, they wake up again. They start dividing. Male and female sexual forms are born. They mate and make oocysts, and the cycle starts over again.

If a person should swallow *Toxoplasma* eggs, either in a speck of soil or in the meat of an infected animal, the parasite will go through this same fast-then-slow progression. Humans hardly know what's happening during a *Toxoplasma* invasion; at worst it feels like a light flu. Once the parasite has retreated to its quiet cyst, a healthy person doesn't notice it at all. It might seem as if *Toxoplasma*, in all its meekness, doesn't warrant mention alongside parasites like trypanosomes and *Plasmodium*. But *Toxoplasma* actually manipulates the immune system of its host as elegantly as these other species do. If the parasite were to multiply madly, grinding up every cell in its host's body, it would find itself inside a corpse rather than a living host. That would hardly be the sort of thing that a cat would want to hunt. *Toxoplasma* wants to keep its intermediate host alive, so it uses its host's immune system to hold itself in check.

Toxoplasma does this with the exact opposite strategy as *Leishmania*. *Leishmania* pushes the immune system to make the T cells that help make antibodies. But *Toxoplasma* releases a molecule that tips the balance in favor of the inflammatory T cells. The inflammatory T cells rise up in huge numbers, turning macrophages into *Toxoplasma* assassins, hunting down the protozoa and blasting them apart. Only *Toxoplasma* that have hunkered down inside tough-walled cysts can survive the attack. From time to time, a few parasites break out of their cysts, squirting a fresh supply of their stimulating molecules, which reenergize the immune system like a booster vaccine. Roused again, the host's macrophages drive the parasites back into their cysts. And so, thanks to *Toxoplasma*'s manipulations, its host stays healthy and able to fight disease while the parasite sits comfortably in its cyst, waiting to reach the promised land of a cat's insides.

Toxoplasma becomes a threat to humans only when the cozy arrangement it creates falls apart. A fetus, for example, doesn't have an immune system of its own. It is protected only by antibodies made by its mother that cross the placenta. The mother's T cells are forbidden from crossing into the fetus, because they would act as if the fetus were a gigantic parasite and would kill it. Maternal antibodies do a good job against a flu virus or *Escherichia coli* bacteria, but they can't protect against *Toxoplasma*. For that, the fetus would need inflammatory T cells to drive them into their cysts. As a result, it's very dangerous for a woman to get a *Toxoplasma* infection during pregnancy. If the parasite manages to pass from her into her fetus, it will reproduce wildly. It will try to make the immune system rein it in, but inside the fetus there's no audience to hear its calls. It simply proliferates until it causes massive, often fatal, brain damage.

In the 1980s, *Toxoplasma* became an accidental killer of another sort of human host: people suffering from AIDS. Human immunodeficiency virus, or HIV, the cause of AIDS, invades inflammatory T cells, using them to reproduce and killing them in the process. When *Toxoplasma* in a person with AIDS pops out of its cyst and divides, it expects a strong immune response to drive it back into hiding. But with hardly any inflammatory T cells left, its host is as helpless as a fetus. The parasite multiplies madly, causing much of its damage in the brain. Its host goes into a delirium and sometimes dies.

For over a decade, doctors could do almost nothing to stop the rampage of *Toxoplasma* in AIDS victims. But in the 1990s, scientists developed drugs that for the first time could slow down the replication of HIV and bring back the inflammatory T cells. In the relative few who can afford these drugs, *Toxoplasma* has gone gladly back into its lair, driven there by a healthy squad of T cells. But the millions who can't afford these drugs continue to face madness brought on by this reluctant parasite.

•   •   •

Surviving the immune system is certainly difficult for a single-celled parasite, but at least it has the advantage of size. It can hide in the pockets of cells or the crooks of lymphatic ducts. The same can't be said for parasitic animals. These multicellular creatures cross the radar of the immune system like vast dirigibles. They are as obvious as a transplanted lung. And without a continual supply of immune-suppressing drugs to hold off the immune system, a transplanted lung will die under its attack. Yet, parasitic animals, some sixty feet long, can live for years inside our bodies, feasting and breeding hundreds of thousands of young.

They thrive because they have many more ways of fooling our immune systems. One remarkable example is the tapeworm *Taenia solium*. Before the eggs of *Taenia* can turn into long ribbons in our bodies, they first need to spend some time in an intermediate host, usually a pig. The pig swallows the eggs with its food, and parasites hatch once they get to the intestines. They use enzymes to dig a hole in the intestines and wriggle their way out. Once they reach a capillary, they ride the bloodstream through the body to a muscle or an organ. There they disembark and settle down, growing into pearly marbles. They can wait for their final host in these cysts for years.

If pigs were the only places where tapeworms spent their cyst years, we'd probably know nothing about how they survive the immune system. But sometimes the eggs of *Taenia solium* end up in humans. (A person with a full-grown tapeworm inside him may get eggs on his hands and then make food for other people, for example.) The eggs proceed to act as if they're in a pig: they hatch, and the larvae go through the same steps of breaking out of the intestines and finding a home somewhere in the body (often the eye or the brain). They then make a cyst, and depending on where they happen to settle, they may be harmless or fatal. If a tapeworm presses against blood vessels, it can kill off tissue; if it causes inflammation in the brain, it can trigger epileptic seizures. If it finds a safer spot, it may go unnoticed for years. But unlike *Toxoplasma*, which essentially falls asleep in its cyst, *Taenia* remains active inside its shell. Through little pores in the cyst wall it sucks in carbohydrates and amino acids, and it grows.

A host's immune system notices the arrival of a tapeworm egg and builds antibodies to it, but by the time it has become organized for an attack, the egg has disappeared; the larva has escaped and formed a cyst for itself. Immune cells crowd around the cyst and build an outer wall of collagen, and yet they can do nothing more. While the cyst takes in food it also releases over a dozen kinds of molecules, each of which stuns the immune system. Complement settles onto the cyst, but the tapeworm releases a chemical that binds to the molecule and stops it from combining into membrane-penetrating drills. The immune cells blast the cyst with highly reactive molecules that can kill tissue, but the tapeworm releases other chemicals that disarm them. And like *Leishmania*, the tapeworms can somehow jam the signals that would normally raise an army of inflammatory T cells. Instead, they encourage the immune system to make antibodies. There's some evidence that suggests why tapeworms would go out of their way to do this. When the antibodies latch onto a cyst, the tapeworm drags them inside its shell and eats them. The tapeworm grows, in other words, by feeding on the futile efforts of the immune system.

Yet, like *Toxoplasma*, the tapeworm doesn't want to kill its intermediate host. It's only when the cyst begins to falter, when it can no longer hold out in the hope of getting into its final host, that it becomes dangerous. The tapeworm can no longer crank out the chemicals it uses to skew the immune system to antibodies. Now the immune system starts making inflammatory T cells tailored to the tapeworm, and they lead the macrophages and other immune cells into action. With such a huge target, the immune cells are worked up into a frenzy. They launch a violent attack that makes the tissue surrounding the cyst swell up, sometimes causing so much pressure that it can kill a person. It isn't the parasite that kills the host, but the host itself.

An even more intimate knowledge of the human immune system can be found in the blood fluke, that passenger from Africa to Australia, that thirty-year-old Methuselah. When young flukes first penetrate the skin, they come to the attention of the immune system. Immune cells manage to kill some flukes early on, perhaps as the parasites struggle through the skin or as they pick their way through the lungs. But having cast off their freshwater coat, the flukes quickly put on a new one that the immune system never quite manages to figure out.

The reason their new coat is so confusing is that it's partially made from the fluke's host. You can see their disguise at work in a simple experiment. When parasitologists take a pair of the parasites out of a mouse and put them in a monkey, the flukes are unharmed and soon start churning out their eggs again. They aren't so lucky if the scientists first inject antigens from mouse blood into the monkey. The injection acts like a vaccine, training the monkey's immune system to recognize and destroy mouse blood antigens. If the flukes are transplanted from the mouse to the vaccinated monkey, the monkey's immune system annihilates them. In other words, the flukes are so much like their mouse host that the monkey's immune system treats them as if they were an organ transplanted from the mouse.

Even though the parasites in this experiment died, it demonstrated a brilliant disguise of theirs. Scientists aren't sure how the flukes cloak themselves, but it seems that their coat is partially made out of the molecules studding our own blood cells. It may be that when the flukes pass by red blood cells or are attacked by white blood cells, they can tear out some of their host's molecules and attach them to their own surface. Thus, to the eyes of the immune system, the parasites are nothing but red shadows in a red river.

These proteins aren't the only things that blood flukes steal from our bodies. Complement molecules settle on the surface of our own cells just as they do on parasites. If they were allowed to go about their business of setting up beacons for macrophages, our immune systems would destroy our own bodies. To avoid this, our cells produce compounds such as decay accelerating factor (or DAF for short), which slices apart the complement molecules. Blood flukes can destroy the complement molecules that land on their own surfaces, and parasitologists have isolated the enzyme that they use. It turns out to be DAF.

It's not clear whether the parasite steals it from its host's cells or owns a gene for making the enyzme. It's possible that at some point in the distant past, a virus that infected humans picked up the gene that makes DAF and then jumped to a blood fluke, adding the borrowed DNA to its new host. In either case, the molecule makes blood flukes as comfortable in our veins as the veins themselves.

In 1995, parasitologists studying blood flukes uncovered a paradox on the shores of Lake Victoria. They were studying Kenyan men who wash cars for a living along the lake. Working in the shallow water, they often get schistosomiasis, the disease caused by blood flukes. The prevalence of AIDS is high in the region as well, so that a fair number of the car-washers had both diseases. HIV destroys inflammatory T cells, the battle-hungry generals that lead macrophages against parasites. As these T cells die off, obscure parasites like *Toxoplasma* rampage through people with AIDS. Yet, blood flukes fare badly alongside HIV. In the Lake Victoria car-washers who had both AIDS and schistosomiasis, the blood flukes shed far fewer eggs than the ones in men who were sick with schistosomiasis alone.

The paradox of the car-washers stems from the fact that blood flukes need to use the human immune system to get their eggs out of their host. Without an immune system, they can't reproduce. Once a female blood fluke lays her eggs in the vein walls, they begin secreting a cocktail of chemicals that manipulates the nearby macrophages. Under the spell of the eggs, the macrophages produce signaling molecules, the most important of which is called tumor necrosis factor alpha (or TNF-α). TNF-α is particularly good at causing inflammation by making the walls of the vein loosen up and by attracting more immune cells. The immune cells try to kill the egg with a spray of poisons, but the egg is protected by its tough shell. All the immune cells can do is wrap themselves around it, weaving an encapsulating shield of collagen.

The immune cells create this capsule (called a granuloma) in the hope of getting rid of the foreign object inside. If a splinter lodges in your thumb, for example, the cells will form a granuloma around it, which will then be carried up to the surface of the skin and be shed from your body. The same thing happens to a granuloma that forms around a fluke egg lodged in the wall of a vein. The granuloma moves through the vein wall and then through the wall of the intestines. This is exactly what the parasite needs to have happen, because it has to get out of its host's body and hatch in water. The parasite, in other words, uses the white blood cells as porters to carry it across an impassable barrier. Once it's on the other side, the immune cells in the granuloma are dissolved in the digestive juices of the intestines, but the tough-shelled egg survives and eventually tumbles out of the body. Hence the paradox of the car-washers of Lake Victoria: AIDS had robbed them of the immune cells the blood flukes needed to send off their young.

It's an elegant way to multiply, but not a very efficient one. The flow of blood in the veins where the blood flukes live travels away from the intestines and up to the liver. As a result, it washes away half of the eggs before they can burrow out. They end up in the liver instead, where they form granulomas. But in the liver, the granulomas can do no good for the parasite, and they can end up killing the host. Parasitologists suspect that the blood flukes may actually keep their damage to their host under control by limiting their own numbers. Like their eggs, adult blood flukes also make the body produce TNF-α. The molecule doesn't do much harm to the adults, but it is lethal to tender young larvae that have just invaded a person but haven't had a chance to build their defenses. As a result, a person who already harbors blood flukes is far less likely to be infected with a new batch. Apparently, the blood flukes help the immune system attack latecomers of their own species to keep the host from getting overcrowded.

What's most impressive about a blood fluke is not how many people it cripples or kills, but how it manages to thrive in the vast majority of its hosts while causing them only a little trouble. They are, in fact, selfish guardians.

• • • •

Only vertebrates have the sort of immune system I've been describing up to this point, with its ever-adapting B and T cells. Invertebrate animals—everything from starfish to lobsters to earthworms to dragonflies to jellyfish—branched away from our own ancestors over 700 million years ago and evolved powerful defenses of their own. Insects, for example, bury intruders in a blanket of cells that ooze out poisons. Eventually the cells form a suffocating seal around the parasite. The parasites that specialize in invertebrates have adapted to their peculiar immune systems, with subterfuges as cunning as anything they use on humans.

One of the best-studied cases is that of the parasitic wasp *Cotesia congregata*. This mosquito-sized wasp uses the tobacco hornworm for its host, a tubby green caterpillar with black hooks on its feet and an orange prong sticking up from its back end like a horn. Scientists have studied this host and parasite so closely because the hornworm is a champion pest, devouring not just tobacco but tomatoes and other vegetables. It is also so big that scientists can simply mash it onto a slide to see what's going on inside.

The attack of a *Cotesia* wasp is so fast you're unlikely to catch it. It lands on a hornworm, crawls up its flank a short way, and stabs its egg-laying syringe into the host. The hornworm may squirm a bit to fight off the wasp, but to no avail. The wasp's eggs hatch inside the hornworm as cigar-shaped larvae. They sip their host's blood while breathing through silvery balloons of tissue on their back ends. The tobacco hornworm has a vibrant immune system, and yet the wasp young go about their business unmolested. But it's not the larvae themselves that stop the immune system. For that, they need a gift from their mother.

The mother wasp injects the eggs as part of a soupy mix. The eggs depend on the soup for their survival: if you take out the eggs, clean off the soup, and then put them directly into a caterpillar, the host's immune system rages full tilt and mummifies the eggs. The parasite survives thanks to millions of viruses swimming in the soup. These viruses are not much like the ones that we're familiar with—the sort that cause a cold, for example. A cold virus wanders from host to host, invading the cells in the lining of the nose and throat, and then commandeering the cell's own proteins in order to make new copies of the virus. Other viruses, like HIV, go so far as to stitch their genes into the DNA of their host cell and make copies of themselves from there. A few go even further: their hosts are born with the virus's DNA already embedded in their own genes and transmit it to their children.

The viruses of parasitic wasps are stranger still. The wasps are born with the virus's genetic code scattered across many of their chromosomes. In males the instructions stay in this scattered form. But as soon as a female begins to take its adult form in her pupa, the virus awakens. In certain cells of her ovary, the pieces of the virus's genome are cut out of the wasp DNA and sewn together, like chapters assembled into a complete viral book. These genes then direct the formation of actual viruses—strands of DNA encased in a protein shell, in other words—and these viruses begin to load up inside the nucleus of the ovary cell. When the nucleus is filled to capacity, the entire cell bursts open, and millions of the viruses float free in the wasp's ovary.

But they don't make a female wasp sick. The wasp actually uses them as a weapon against the tobacco hornworm. When it injects the viruses into a caterpillar along with its eggs, the viruses start invading the host's cells in a matter of minutes. They commandeer the host's DNA, forcing the cells to make strange new proteins normally never seen inside a hornworm, which flood the body cavity of the caterpillar. These proteins destroy the hornworm's immune system. The cells start sticking to one another instead of to the parasites, and then they burst open. The host is left as immunologically helpless as a person with full-blown AIDS (which is also caused by a virus that blows apart immune cells). Thanks to the virus, the wasp eggs can hatch and begin to grow without any harrassment by their host.

But unlike a person infected with AIDS, the hornworm recovers from the wasp virus after a few days. By then, the wasp larvae seem to be able to handle the immune system on their own, without help from mother. They may fool their host in ways similar to the ways blood flukes fool us, by borrowing the insect's own proteins or by mimicking them.

It may seem perverse for a virus to do the dirty work for another organism, even going so far as wiping out a host's immune system only to be wiped out itself. But within every egg that the virus protects, there are instructions for making new viruses that will survive if some viruses attack the host. At the same time, though, it may be wrong to think of a virus as a separate organism with its own evolutionary ends. The truth may be even more perverse, for the virus's DNA resembles some of the wasp's own genes. The resemblance may actually be hereditary: the virus may descend from a fragment of wasp DNA that mutated into a form that escaped from the normal way genes are copied and stored. It may not be strictly correct to call the viruses viruses at all—they may represent a new way that wasps package their own DNA. (One scientist has suggested calling the viruses genetic secretions.) If that's the case, then parasitic wasps are managing to insert their own genes into another animal's cells to make it a better place for the wasps' to live.

These wasps may seem as if they belong on another planet, but they actually demonstrate a universal quality to parasites here on Earth: parasites find ways to battle immune systems, tailored precisely to the peculiarities of their host. Whether they end up killing or sparing their hosts depends on how they can best make more of themselves.

4

A Precise Horror

You still don't know what you're dealing with, do you? Perfect organism. Its structural perfection is matched only by its hostility . . . I admire its purity; unclouded by conscience, remorse, or delusions of morality.

—Ash to Ripley in *Alien* (1979)

Ray Lankester had nothing but contempt for *Sacculina*, the barnacle that degenerates practically into a plant. He was appalled by the way it had clambered down the ladder of evolution, a symbol of all things backward and lazy. Strange, then, that *Sacculina* now turns out to be an emblem for just how sophisticated a parasite can get.

Lankester's mistake didn't stem simply from a loathing for all parasites; biologists of his day just didn't know much about *Sacculina*. It's true that these parasites start life as free-swimming larvae. Through a microscope they look like teardrops equipped with fluttering legs and a pair of dark eyespots. Biologists in Lankester's day thought *Sacculina* was a hermaphrodite, but in fact, it comes in two sexes. The female larva is the first to colonize a crab. She has sense organs on her legs that can catch the scent of a host, and she will dance through the water until she lands on its armor. She crawls along an arm as the crab twitches in irritation or perhaps the crustacean equivalent of panic. She comes to a joint on the arm, where the hard exoskeleton bends at a soft chink. There she looks for the small hairs that sprout out of the crab's arm, each anchored in its own hole. She jabs a long hollow dagger through one of the holes, and through it she squirts a blob made up of a few cells. The injection, which takes only a few seconds, is a variation on the moulting that crustaceans and insects go through in order to grow. A cicada sitting on a tree separates a thin outer husk from the rest of its body, and then pushes its way out of the shell. It emerges with a new exoskeleton that stays soft long enough to stretch as the insect goes through a growth spurt. In the case of the female *Sacculina*, however, most of her body becomes the husk that is left behind. The part that lives on looks less like a barnacle than a microscopic slug.

The slug (whose existence was discovered only in 1995) plunges into the depth of the crab. In time it settles in the crab's underside and grows, forming a bulge in its shell and sprouting the roots that so appalled Lankester. Biologists still call these things roots, but they are hardly like what you find under a tree. Fine fleshy fingers cover them, much like the ones lining our own intestines or the skin of a tapeworm. Unlike the exoskeleton of a regular crustacean, it is never moulted. Instead, the roots draw in nutrients dissolved in the crab's blood. The crab stays alive during this entire time; you can't tell it apart from healthy crabs as it wanders through the surf, eating clams and mussels. Its immune system can't fight off *Sacculina*, and yet it can go on with its life with the parasite filling its entire body, the roots even wrapping around its eyestalks.

The female *Sacculina*'s bulge grows into a knob. Its outer

layer chips away, slowly revealing a portal at the top. She will remain at this stage for the rest of her life unless a male larva finds her. He lands on the crab and walks along its body until he reaches the knob. At its summit, he finds the pin-sized opening. It's too small for him to fit into, and so, like the female before him, he moults off most of himself, injecting a vestige of it into the hole. This male cargo—a spiny, reddish brown torpedo a hundred-thousandth of an inch long—slips into a pulsing, throbbing canal, which carries him deep into the female's body. He casts off his spiny coat as he goes, and in ten hours he ends up at the bottom of the canal. There he fuses to the female and begins making sperm. There are two of these wells in each female *Sacculina*, and she typically carries two males with her for her entire life. They endlessly fertilize her eggs, and every few weeks she produces thousands of new *Sacculina* larvae.

The crab begins to change into a new sort of creature, one that exists to serve the parasite. It can no longer do the things that would get in the way of *Sacculina*'s growth. It stops moulting and growing, which would funnel away energy from the parasite. Crabs can typically escape from predators by severing a claw and regrowing it later on. Crabs carrying *Sacculina* can lose a claw, but they can't grow a new one in its place. And while other crabs mate and produce a new generation, parasitized crabs simply go on eating and eating. They have been spayed. The parasite is responsible for all these changes.

Despite being castrated, the crab doesn't lose its urge to nurture. It simply directs its affection toward the parasite. A healthy female crab carries her fertilized eggs in a brood pouch on her underside, and as her eggs mature she carefully grooms the pouch, scraping away algae and fungi. When the crab larvae hatch and need to escape, their mother finds a high rock on which to stand, and she bobs up and down to release them from the pouch into the ocean current, waving her claws to stir up more flow. The knob that *Sacculina* forms on a crab sits exactly where the brood pouch would be, and the crab treats the parasite knob as if it were its own pouch. She strokes it clean as the larvae grow, and when they are ready to emerge, she forces them out in pulses, shooting out heavy clouds of parasites. As they come spraying from her body she waves her claws to help them on their way. Male crabs aren't out of reach from *Sacculina's* powers, either. Males normally develop a narrow abdomen, but infected males grow abdomens as wide as females, wide enough to accommodate a brood pouch or a *Sacculina* knob. A male crab even acts as if he has the female's brood pouch, grooming it as the parasite larvae grow and bobbing in the waves to release them.

Simply living within another organism—locating it, traveling through it, finding food and a mate inside, altering the cells that surround it, outwitting its defenses—is a tremendous evolutionary accomplishment. But parasites such as *Sacculina* do more: they control their hosts, becoming in effect their new brain, and turning them into new creatures. It is as if the host itself is simply a puppet, and the parasite is the hand inside.

This puppetry takes different forms depending on the particular parasite and what it needs from its host at its particular stage of life. When a parasite has first settled into a comfortable spot in its host, food is the first order of business. Once a tobacco hornworm has been rendered defenseless by the viruses of the parasitic wasp *Cotesia congregata*, the wasp's eggs are ready to hatch and grow. Rather than just passively soak up the food around it, the wasp changes the way its host eats and digests its food. The more wasps in a given host, the bigger the host will grow—up to twice its normal size. And once the caterpillar eats a leaf, the wasps alter the way it breaks it down. Normally a hornworm would convert a lot of the leaf into fat, a stable form of energy that it can store away for the time when it will fast inside its cocoon. But once it is infected by wasps, the hornworm turns its food into sugar, a quick source of energy that the parasites use for fast growth.

A parasite lives in a delicate competition with its host for

A Precise Horror

the host's own flesh and blood. Any energy that the host uses itself could go instead to the growing parasite. Yet, a parasite would be foolish to cut off the energy to a vital organ like the brain, since the host would no longer be able to find any food at all. So the parasite cuts off the less essential things. As *Cotesia congregata* robs the caterpillar of its fat stores it also shuts down its host's sex organs. Male caterpillars are born with big testes, and normally they channel a lot of the energy from their food into building them up even more. When a parasitic wasp lives inside the male, however, the testes shrivel up. Castration is a strategy that any number of parasites have hit on independently—*Sacculina* does it to crabs, and blood flukes do it to the snails they invade. Unable to waste energy on building eggs or testes, on finding a mate, or on raising young, a host becomes, genetically speaking, a zombie: one of the undead serving a master.

Even flowers can become zombies to their parasites. A fungus called *Puccinia monoica* lives inside mustard plants that grow on the slopes of Colorado mountains. The fungus sends its tendrils throughout the stem of the mustard plant, feeding on the nutrients the flower draws from the sky and the soil. In order to reproduce, it needs to have sex with the *Puccinia* inside another mustard plant. To do so, the fungus stops the plant from sending up its own delicate little flowers and forces it to turn clusters of its leaves into brilliant yellow imitations of flowers. These fakes look exactly like other flowers found on the mountains, not just in visible light but in ultraviolet light as well. They lure bees, which can feed on a sweet, sticky substance that the fungus forces the plant to produce on the imitation flowers. The fungus crams its sperm and its female sex organs into them, so that the bees can fertilize the fungus as they travel from mustard plant to mustard plant. But the plant itself remains sterile.

No matter how comfortable a parasite may make itself by altering its host, it has to leave sooner or later. Some parasites head on to the next host in their life cycle, others go to a free-living adulthood, and in many cases the parasites stage-manage a careful exit. Simply letting the host go on with its normal life would mean death for most parasites. The tobacco hornworm normally moults five times and then wanders down from its plant to the ground. It digs a few inches into soil and forms its cocoon, where it stays until it emerges as a moth. When hornworms are parasitized by the wasp *Cotesia congregata*, however, they take a different path. They moult only twice, and they never get the call to wander off their plant. Instead, they go on chewing leaves, nurturing their parasites until the wasps are ready to emerge. The hornworm then slows down and stops eating, losing its appetite. The wasps seem to be responsible for the anorexia, because a healthy hornworm will happily devour dozens of wasp cocoons.

Another species of wasp goes even further, turning its host—the cabbage worm caterpillar—into a bodyguard. When the wasp's larvae have matured, they paralyze the cabbage worm and push their way out of its abdomen. They then spin their cocoons on the underlying leaf. Yet, even after the wasps have devoured the guts of the caterpillar and riddled it with escape hatches, the cabbage worm recovers. It doesn't limp away; instead, it weaves a mesh over the wasps to shield them from other parasites and coils itself on top. If anything should disturb the caterpillar as it stands guard, it lashes out, biting and spitting up noxious liquids—in other words, protecting the cocoons. Only when the wasps emerge from their cocoons does the cabbage worm end its duty to them and lie down to die.

While wasps can live on dry land once they've left their hosts, many other parasites need to get to water. There are parasitic nematodes, for instance, that live as free-living adults in streams, where they mate and lay their eggs. When their offspring hatch, they attack the mayfly larvae that live alongside them. The nematodes pierce through the mayfly's exoskeleton and curl up inside its body cavity. There they grow as the mayfly grows, siphoning off its food. The mayflies go through a long, lingering insect adolescence in the water before they transform into delicate, long-winged forms. The males rise from the water and form great clouds that attract the females. The nematodes rise invisibly into the cloud within their hosts.

Male and female mayflies find each other in the swarm. Embracing, they fall to the grasses and reeds along the stream, and mate. You can tell the difference between the sexes not only by their genitals (the males have little claspers to help them mate) but by other parts of their bodies such as their eyes: the female has small eyes pointing out to either side, while those of the male bulge out so much that they touch over the top of its head. Once they've mated, the males have finished their life's work. They fly lazily away from the stream to find a place to die. The females, meanwhile, make their way upstream to find a protruding rock. They crawl under it and bob their abdomens up and down as they lay their eggs. If the female is carrying a nematode, the full-grown parasite breaks out of the mayfly's abdomen and burrows away into the gravel to find a mate of its own, leaving its host dead.

The nematode's strategy has one big, obvious flaw: if it happens to climb inside a male mayfly, it will end up in a patch of grass. Instead of getting back to the water, it will die with its host. The nematode has a solution, one that's reminiscent of *Sacculina:* it turns the male into a quasi-female. When an infected male mayfly matures, he never forms his claspered genitals or even his high-domed eyes. The nematode makes him not only look like a female but act like one, too. Instead of flying away, he drops down to the stream, even going so far as to try to lay imaginary eggs as the parasite bursts out of his body.

The nematode needs to get back to the stream for two reasons—to move on to the next stage of its life, and to be in a place where its offspring will be able to find a mayfly of their own to invade. Getting to the next host is a consuming passion among parasites, because there is no alternative: "Live free and die" is their motto. A fungus that lives inside house flies provides a spectacular example of this. When the spores of the fungus make contact with a fly, they stick to its body and dig tendrils into the fly's body. The fungus spreads throughout the fly's body with *Sacculina*-like roots and sucks up the nutrients of its blood, making the fly's abdomen swell as it grows. For a few days the fly lives on normally, flying from spilled soda to cow turd, using its proboscis to sponge up food. But sooner or later it gets an uncontrollable urge to find a high place, be it a blade of grass or the top of a screen door. It sticks out its proboscis but uses it as a clamp this time, gluing itself to its high perch.

The fly lowers its front legs, tilting its abdomen away from the surface. It flaps its wings for a few minutes before locking them upright. The fungus has meanwhile pushed its tendrils out of the fly's legs and belly. On the tips of the tendrils are little spring-loaded packages of spores. In this bizarre position, the fly dies, and the fungus catapults out of its corpse. Every detail of this death pose—the height, the angles of the wings and the abdomen—all put the fungus in a good position for firing its spores into the wind, to shower down on flies below.

As if this were not enough of an accomplishment for a speck of fungus, infected flies always die in this dramatic way just before sunset. If the fungus matures to the point where it can make spores in the middle of the night, it doesn't: it holds off the process, waiting through the dawn and the day. It is the fungus, not the fly, that decides not only how it will die but *when*—just before sundown. Only then is the air cool and dewy enough for the spores to develop quickly on another fly, and only then are healthy flies leaving the air for the night and moving down toward the ground, where they make easy targets.

Parasites such as this fungus use their hosts to get to other hosts of the same species. But for many other parasites, the game is more complicated: they have to make their way though a whole series of different animals. Sometimes they force their current host to get into the vicinity of their next one. Along the coasts of Delaware lives a fluke that uses mud snails as its first host and fiddler crabs as its second. The only problem is that the snails live in the water and the crabs live on shore. But when the snails are infected by the fluke, they change their behavior. They grow restless; they wander onshore or onto sandbars during low tides and linger there while healthy snails keep to the water. They shed their flukes on the sand, putting the parasites so close to the fiddler crabs that they can easily burrow into them. It's as simple as getting a taxi to a bus station.

Another species of fluke can be found in the meadows of Europe and Asia, along with a few in North America and Australia. Known as *Dicrocoelium dendriticum*, or the lancet fluke, it makes cows and other grazers its host as an adult, and the cows spread their eggs in their manure. Hungry snails swallow the eggs, which hatch in their intestines. They drill through the wall of a snail's gut and settle in the digestive gland. There the flukes produce a generation of cercariae, which make their way to the snail's surface. The snail tries to defend itself from the parasites by blocking them off with walls of slime. The slime balls up around the cercariae, which the snail coughs up and leaves behind in the grass.

Next, along comes an ant. To an ant, a slime ball is positively delicious. Along with the slime, the ant may also swallow hundreds of lancet flukes as well. The parasites slide down into its gut, and they then wander for a while through its body, eventually moving to the cluster of nerves that control the ant's mandibles. The parasites all travel together on this trip, but after visiting the nerves, they split up. Most of the lancet flukes head back to the abdomen, where they form cysts, but one or two stay behind in the ant's head.

There they do some parasitic voodoo on their hosts. As the evening approaches and the air cools, the ants find themselves drawn away from their fellow ants on the ground and upward to the top of a blade of grass. Like flies infected with a fungus, the ants clamp down on the tip of the grass. But the lancet fluke has a different goal than the fungus does. The fungus uses its host as a catapult to shower its spores on other insects. The lancet fluke can continue to live only if it can get inside its final host, a mammal. Clamped to the tip of a grass blade, the infected ant is likely to be devoured by a cow or some other grazer passing by. When the ant tumbles into the cow's stomach, the flukes burst out and make their way to the cow's liver, where the flukes will live as adults.

But the lancet fluke, like the fungus, is very aware of the passing of time. If the ant sits the whole night without being eaten and the sun rises, the fluke lets the ant loosen its grip on the grass. The ant scurries back down to the ground and spends the day acting like a regular insect again. If the host were to bake in the heat of the direct sun, the parasite would die with it. When evening comes again, it sends the ant back up a blade of grass for another try.

Most parasites rarely try this sort of thing on humans, but a few do it very well. The guinea worm spends its early life curled up inside a copepod swimming in water. A person drinking that water swallows the copepod, and when it dissolves away in stomach acid, the guinea worm escapes. It slips into the intestines and burrows out into the abdominal cavity. From there it wanders through the connective tissue until it finds a mate. The two-inch male and the two-foot female have sex, and then the male looks for a place to die. The female slithers through the skin until she reaches a leg. As she travels, her fertilized eggs begin to develop, and by the time she has reached her destination the eggs have hatched and become a crowd of bustling juveniles in her uterus.

These juveniles need to get into a copepod if they are to become adults themselves, and so they drive their human host to water. They press against their mother's uterus so hard that they force it partially out of her body, letting some of the larvae spill out. Adult guinea worms tame the human immune system so that they can travel through our bodies unharmed, but the juveniles do just the opposite. They draw a quick reaction that brings immune cells rushing to them, making the skin around them swell and blister. The easiest way for a victim to get some relief from the hot pain of the wound is to pour cool water on it or just stick the leg in a pond. The juveniles that have already escaped their mother inside the blister respond to the splash by swimming free. The mother responds to the water as well by getting rid of more of her young. She doesn't herniate herself the way she did before; this time she lets her babies escape through an even stranger route: her mouth. For every splash, half a million baby guinea worms come heaving up through her esophagous. The contractions pull her out of the wound bit by bit until she and her young have all left the host—the mother to die, the young to search the water for a new copepod to curl up inside.

This manipulation works best when humans and copepods all depend on scarce supplies of water, because that makes it more likely a person will dump guinea worm larvae where their next host can be found. Not surprisingly, dracunculiasis, the disease caused by the guinea worm, is particularly bad in deserts, where people crowd around oases.

The guinea worm is the sort of parasite that is content to sit in its first host until it is accidentally swallowed by its next one. Other parasites don't rely so much on luck. Their hosts come into regular contact, usually to eat or be eaten. Biting insects seek out humans and other vertebrates and drink their blood, and they are—not coincidentally—filled with parasites trying to get into us. Malaria and filariasis are spread by mosquitoes, sleeping sickness by tsetse flies, kala-azar by sand flies, river blindness by black flies. (Bacteria and viruses come along for the ride as well, spreading bubonic plague, dengue fever, and other diseases.) These parasites swim into the wound made by the insect and then live in our skin or bloodstream, where they are likely to be taken in the bite of the next passing insect. But simply being in the right place is not enough for many of them—they change the behavior of the insects to make them spread the parasites faster.

Drinking blood is not easy. When a mosquito lands on your arm, it has to drive its proboscis through the tough outer layers of your skin and then snake it around for a while to find a blood vessel. The longer it takes, the better its chances of getting slapped and being reduced to a bloody smear. And once the mosquito hits blood, your body responds by clotting the wound. Platelets swarm around the mosquito's proboscis, releasing chemicals that make them form sticky clumps and attract other platelets. As the mosquito tries to drink, its smooth cocktail of blood turns into a thick milk shake. To buy themselves more time, mosquitoes put chemicals in their saliva that fight against the clotting. One of them, apyrase, cuts apart the glue made by the platelets; other chemicals widen blood vessels to bring in more blood.

The risks of drinking blood make mosquitoes afraid of commitment. If they find it too difficult to draw blood from a host, they'll quickly fly to a new patch of skin. But if that host has malaria, the parasites inside will make him more attractive. Malaria interferes with the platelets of its host, making them do a bad job of clotting. When a mosquito hits blood in a person with malaria, it will find it easier to drink and will be more likely to suck it up, and the parasite along with it.

Once it gets into a mosquito, *Plasmodium* needs time before it can travel into another human. It needs to move into the mosquito's gut, mate with other *Plasmodium* parasites, and reproduce. More than ten thousand ookinetes are formed this way in ten days. They develop into sporozoites that migrate up to the salivary gland, where they're finally ready to enter a human. But up to that point, it doesn't do the parasite any good for the mosquito to eat. The risks of getting squashed in mid-bite are offset by no benefit. So *Plasmodium* does its best to discourage its host from eating. A mosquito with ookinetes in it will give up trying to take a blood meal more easily than a parasite-free one.

Once the parasite has reached the mosquito's mouth, though, it wants the mosquito to start biting as much as possible. *Plasmodium* travels to the salivary glands, homing in on a lobe that is responsible for making the anticoagulant molecule apyrase. There it proceeds to cut off the mosquito's apyrase supply, so that when the insect drives its proboscis into a new host, it has a harder time keeping the blood flowing. It has to visit more hosts to drink the same amount of blood. At the same time, *Plasmodium* makes the mosquito hungrier, drinking more blood and visiting more hosts to get it. As a result, a sick mosquito is twice as likely as a healthy one to drink the blood of two people in a night. The sick mosquito, carrying more blood to more hosts, becomes a far more effective way to spread malaria.

Plasmodium makes a predator—a mosquito—come into contact with its prey—us. Parasites can use the opposite arrangement as well, by living first in prey and waiting until a predator eats it. Some parasites are willing to sit and wait for their intermediate host to be devoured. But many are not so patient. A fluke called *Leucochloridium paradoxum* makes snails its first host, but makes insect-eating birds its final host, even though the birds have no appetite for snails. The flukes get the bird's attention by pushing their way into the eye tentacles of the snail. Covered in brown or green stripes, the parasites are visible through the transparent tentacles, and to a bird they look like caterpillars. A bird attacks the snail and ends up with nothing but a bellyful of parasites.

Other parasites can change their host's skin to become a more obvious target. Some species of tapeworms live in the guts of the threespine stickleback fish for a few weeks, and when they want to get into a bird, they turn the fish orange or white. They can also alter the behavior of the fish to get the attention of the birds. Normally, sticklebacks keep diligently away from the water birds that like to eat them. They try to stay well below the water's surface, and if a heron should stick its head underwater, they will dart away, passing up the opportunity to eat. But when they are infected by tapeworms, they become buoyant so that they can't help but swim near the surface, and they become fearless as well, chasing after food even if a bird is dangerously close by.

Sometimes it's not enough for a parasite to make its host vulnerable to attack; sometimes it sends its host straight into harm's path. Such is the case with thorny-headed worms. Many species of these parasites start off inside invertebrates that live in lakes and rivers. They then become adults in birds, where they drive their barbed heads deep into the lining of the intestines. A small crustacean named *Gammarus lacustris* feeds near the surface of ponds and rivers, but as soon as its predator—a duck—comes around, it escapes by diving away from the light and thus down to the bottom of the water. When a thorny-headed worm gets inside a *Gammarus*, though, it does the exact opposite. If a duck comes on the scene, *Gammarus* feels an unshakable attraction toward light—and thus moves up to the surface of the water. When it reaches the surface, it skims along until it finds a rock or a plant. Once it makes contact, it clamps its mouth down, practically offering itself up to the duck.

Toxoplasma, the protozoan lodged in billions of human brains, may seem like a gentle creature that wouldn't get involved in mind control. After all, it hides safely in its cysts and declines to kill its hosts. But its tameness is only part of its unconscious calculation of how to boost its odds of getting into its final host. *Toxoplasma* needs to move between cats and their prey and back to complete its life cycle, and a dead rat won't attract many cats. But *Toxoplasma*, it turns out, does what it can to help the cats kill their prey.

For several years scientists at Oxford University have been studying the effects of *Toxoplasma* on the behavior of rats. They built a six-foot by six-foot outdoor enclosure and used bricks to turn it into a maze of paths and cells. In each corner of the enclosure they put a nest box along with a bowl of food and water. On each nest they added a few drops of a particular odor. On one they added the scent of fresh straw bedding, on another the bedding from a rat's nest, on another the scent of rabbit urine, on another the urine of a cat. When they set healthy rats loose in the enclosure, the animals rooted around curiously and investigated the nests. But when they came across the cat odor, they shied away and never returned to that corner. This was no surprise: the odor of a cat triggers a sudden shift in the chemistry of rat brains that brings on intense anxiety. (When researchers test anti-anxiety drugs on rats, they use a whiff of cat urine to make them panic.) The anxiety attack made the healthy rats shy away from the odor and in general made them leery of investigating new things. Better to lie low and stay alive.

Then the researchers put *Toxoplasma*-carrying rats in the enclosure. Rats carrying the parasite are for the most part indistinguishable from healthy ones. They can compete for mates just as well and have no trouble feeding themselves. The only difference, the researchers found, is that they are more likely to get themselves killed. The scent of a cat in the enclosure didn't make them anxious, and they went about their business as if nothing was bothering them. They would explore around the odor at least as often as they did anywhere else in the enclosure. In some cases, they even took a special interest in the spot and came back to it over and over again.

By turning rats into rodent kamikazes, *Toxoplasma* probably increases its chances of getting into cats. If it makes the mistake of getting into a human instead of a rat, it has little hope of making that journey, but there's some evidence that it still tries to manipulate its host. Psychologists have found that *Toxoplasma* changes the personality of its human hosts, bringing different shifts to men and women. Men become less willing to submit to the moral standards of a community, less worried about being punished for breaking society's rules, more distrustful of other people. Women become more outgoing and warmhearted. Both changes seem to break down the fear that might keep a host out of danger. They're hardly enough to make people throw themselves at lions, but they're a very personal reminder of the ways in which parasites try to take control of their destiny.

Scientists have known about these sorts of transformations for more than seventy years, but they didn't think they were actually manipulations. Parasites couldn't possibly mastermind pinpoint changes to their plainly superior hosts. They could only cause random kinds of harm, and maybe by chance the damage altered their host. Only in the 1960s did scientists begin to think seriously about the possibility that a parasite might be able to engineer the physiology of its host, or even its behavior. And thereupon emerged a long line of cases that seemed, on their faces, to be just that.

Most of the cases came from eukaryote parasites, although certainly bacteria and viruses can act as puppet-masters sometimes. A sneeze carries away cold viruses to new hosts; the Ebola virus seems to take advantage of our respect for the dying and the dead by making its victims gush blood, which gets on the bodies of people handling their bodies, infecting them as well. But if you look over the documented cases of manipulators, bacteria and viruses make up a tiny portion. It may be that their needs are pretty simple: they rarely need to use more than one species as a host, and they can just ride along during the regular contacts between hosts—be it sex, a handshake, or the bite of a tick. There may in fact be a lot of manipulators waiting to be revealed among bacteria and viruses. They may still be hidden, thanks to the fact that most people who study viruses and bacteria primarily think in terms of diseases, symptoms, and cures. They tend not to think like parasitologists, who treat their subjects more as living beings that have to survive in their hosts and get to new ones.

The great danger in studying parasite manipulations is to see cunning strategies of parasites where none exist. Some changes to a host can be simple damage. And if a person can tell that a parasite has changed the color of a fish, that doesn't really mean anything. What matters is whether the change actually makes it easier for a bird to eat it. The only way to demonstrate that a manipulation is genuine is to run experiments, and the first ones that demonstrated real manipulations with significant effects were performed in the 1980s by Janice Moore, a parasitologist at Colorado State University. Her parasites of choice were a species of thorny-headed worms that live as larvae inside pill bugs on the forest floor, live as adults in starlings, and pass their eggs out in the bird droppings for more pill bugs to pick up.

Moore built chambers out of Pyrex pie plates to measure the behavior of the infected pill bugs. In one experiment, she wanted to see how the pill bugs responded to humidity. She set one plate on top of another to create an enclosed space. Then she divided the space into two chambers with a glass barrier, leaving only a narrow slit between them, which she covered with a piece of nylon mesh. She made one of the chambers humid by pouring into it potassium dichromate—a chemical that reacts with air to make water. In the other side she poured salt water, which made the air dry by pulling water out of it. She then let a few dozen pill bugs loose inside the pie plate house she had built, and waited to see which chamber, humid or dry, they chose. Afterward, she dissected them and looked inside to see whether they carried the larvae of thorny-headed worms.

In another experiment, she built a little shelter for the pill bugs with a tile sitting on top of four pebbles in the middle of a pie plate. She watched to see whether they hid under it or walked out in the open. And in a third one, she poured colored gravel into a pie plate—one half white, the other black—to see whether pill bugs were drawn to light or dark backgrounds.

Pill bugs live in moist forest soils, where they can hide from the birds that would eat them. If you take them out, they'll scurry back in. They're attracted to the soil by factors like humidity, dim light, and dark colors. The healthy pill bugs that

Moore studied behaved this way in her pie plates. They stayed in the humid chamber and avoided the dry one; they hid under the shelter she made for them; and they chose dark gravel over light. But the pill bugs that carried thorny-headed worms could be found wandering into the dry part of her chamber much more often than the healthy ones. A parasite would make its host crawl over the white gravel more often, and be far less likely to hide under the shelter. The parasitized pill bugs could no longer recognize these vital clues, and they became easier prey for birds.

But rather than imagine what might make a bird's life easier, Moore let the birds tell her themselves. She let pill bugs roam around a cage in which she kept starlings. The birds ate the pill bugs, and she found that they preferred the infected ones over the healthy ones. In another experiment, she set up nest boxes for starlings, which came and raised nestlings in them. They would hunt in the surrounding fields for food—including pill bugs—and bring it back to the box. Moore loosely tied pipe cleaners around the necks of the nestlings, closing off their throats just enough so they couldn't swallow their meals. By picking through their mouths and the nest, Moore could collect the pill bugs the adult birds had brought. She dissected them to check for parasites and found that the parasitized pill bugs turned up in the nests far more often than they should have. At a typical site, fewer than 1 percent of the pill bugs carried the thorny-headed worms, but 30 percent of the ones Moore collected from the nestlings were infected.

Moore's experiments were followed by other careful tests, and in many cases the parasites in question did indeed boost their success by altering their hosts. Once parasitologists showed that these manipulations were real, they began to ask how exactly the parasites manage them. Each parasite probably uses its own special mechanism, some of which may be pretty simple. When tapeworms grow inside three-spined stickle-backs, filling their entire body cavity and soaking up most of the food their hosts eat, they probably make the fish ravenous. Their hunger pushes the sticklebacks to take more risks to get food, not to dart away when they realize a bird is nearby. To the tapeworm, danger means deliverance.

More often, though, the mechanisms are far more sophisticated. Parasites have mastered the vocabulary of their hosts' neurotransmitters and hormones. Parasitologists are pretty confident that this is the case, even though they haven't yet found a particular molecule that they know can alter a host in a particular way. The bodies and brains of animals are just too noisy with the traffic of signals for scientists to catch a quick transmission from parasites. But parasitologists can still say a lot about those parasitic molecules indirectly, in the same way you can judge a man by his shadow.

Recall for a moment poor *Gammarus*, sent hurtling up to the surface of a pond by a thorny-headed worm, where it clamps down on a rock until a duck eats it. Clearly, something is wrong with its nervous system, because the same sensation that would send a healthy *Gammarus* to a river bottom produces the opposite reaction in a sick one. Biologists have pulled out the neurons of *Gammarus* infected with thorny-headed worms. They've stained them with compounds that make the neurons light up if they carry certain neurotransmitters. When they've looked for a neutrotransmitter called serotonin, the neurons have lit up like Christmas trees.

You can find serotonin in just about any animal you look at. In humans and other mammals, it seems to stabilize the brain. When levels of serotonin drop, people may become obsessive, depressed, violent. (Prozac is designed to counter depression by boosting serotonin.) Serotonin also plays a role in invertebrate brains, although scientists aren't sure what that role is. They do know that something interesting happens when they inject serotonin into *Gammarus*. If a healthy *Gammarus* gets a shot, it will often try to grab on to something and hold tight.

Why should serotonin cause *Gammarus* to cling? It may

have something to do with sex. When *Gammarus* mate, the male grabs the female with his legs and pulls his abdomen down toward hers. He will ride her for days, waiting for her to moult. When she does, she puts her eggs in a pouch under her belly. The male fertilizes the eggs and continues to hold on, guarding her against other males that want to mate.

The mating male's pose is exactly like the one that thorny-headed worms force *Gammarus* to take. And if parasitologists inject a drug into infected *Gammarus* that blocks the effects of serotonin, they stop clinging for a few hours. It may be that the thorny-headed worm secretes a serotonin-boosting molecule. The parasite may trigger a sequence of signals that makes the *Gammarus* think it's having sex, even making the females take on the male's role in the mating.

When parasitologists figure out the full story of parasitic manipulators, it will turn out to be more sophisticated than this. It's unlikely that parasites use a single molecule to control their hosts; they come equipped with a big pharmacy full of drugs ready to be dispensed at different times in the parasite's life when it needs different things. That's the picture that emerges when scientists have pooled their efforts to study the full cycle of one particular parasite, such as the tapeworm *Hymenolepis diminuta*. *Hymenolepis* adults live and mate inside the bowels of rats, where they grow to be a foot and a half long. Their eggs end up in rat droppings, which are regularly devoured by beetles. Once inside a beetle, the tapeworm's egg membrane dissolves away, revealing a spherical creature with three pairs of hooks. It uses those hooks to claw out of the beetle's gut and into its circulatory system, where it grows in a little over a week into a short-tailed form. There it waits for the beetle to be eaten by a rat, where it will take its final adult form. The whole cycle often takes place in grain silos or flour warehouses, where the beetles devour the food, the rats eat the beetles, and then the rats leave their droppings in the grain.

A Precise Horror

The tapeworms begin manipulating the beetles even before they are inside them. Beetles are lured to egg-laden droppings by an aroma that's apparently irresistible to the insects. If a beetle should come across droppings from a healthy rat and droppings from a parasitized one, it's more likely to choose the pile with the tapeworm eggs. If you trap the fragrance of infected dung and preserve it in liquid, a drop of this perfume will bring beetles scurrying. No one knows if the eggs themselves produce the scent, or if it's one of the chemicals produced by the adult tapeworms inside the rats, or if the parasites somehow change that rat's digestion so that the host itself makes it. Whichever is the case, it's enough to seduce the beetles into eating a tapeworm, perhaps into being eaten by a rat.

Once inside the beetle, the tapeworm then uses more chemicals to sterilize it. Like most other insects, a beetle stores up reserves of energy in a structure called the fat body that runs along its back. Female beetles use some of this material to build the yolks for their eggs. To get the reserves to the eggs, they have to send a hormone signal to the fat body. The fat body cells respond to it by making a yolk ingredient called vitellogenin. The vitellogenin leaves the fat body and flows through the beetle until it reaches the eggs in the ovaries. A beetle egg is surrounded by a retinue of helper cells that leave only a few cracks between them. The cracks are so few and so small, in fact, that it's hard for anything to get through them and to the egg itself. But when the right hormones latch onto these helper cells, they make them shrink, opening up the spaces. With enough of these hormones, the vitellogenin can reach the egg itself and turn into yolk.

The tapeworm can destroy this chain of events at several links. It makes a molecule that gets into the fat body and slows down the cells as they make vitellogenin. Some vitellogenin still gets out of the fat body, but little of it seems to reach an egg. It appears that the tapeworm makes yet another molecule that can lock into the receptors on the helper cells in the ovaries. It plugs up the receptors to stop the hormone from latching on and making the helper cells shrink. The helper cells stay swollen, so the vitellogenin can't get into the egg. The effect of these molecules is to stop the beetle from diverting what could be perfectly good tapeworm food into its own eggs.

Once it has matured inside the beetle, the tapeworm is ready to find itself a rat. The beetle certainly wouldn't agree, so the parasite has to pull open another drawer of drugs. Some of them—probably opiates that blunt feelings of pain and fear—make the beetle less conscientious about concealing itself. Put it on a pile of flour, and the beetle will be likely to wander the surface instead of burrowing out of sight. The tapeworm makes it sluggish, slow to escape from an attack. Still, an infected beetle does its best to defend itself if a rat should take it in its jaws. A flour beetle comes equipped with a pair of glands on its abdomen that it uses to release a foul-tasting chemical, and a rat that grabs the beetle in its mouth is likely to spit it out. But once the tapeworm reaches maturity, it blocks the gland from making its poison. When the infected beetle tries to defend itself, it doesn't taste all that bad to the rat; it is thus far more likely to be eaten than a healthy counterpart. From beginning to end, the beetle is guided and tugged by its parasite.

●   ●   ●

If you turn off the Ventura Freeway at the town of Carpinteria, California, and drive a short way toward the ocean, passing a teddy bear warehouse and a set of train tracks, you come to a chain link fence. Beyond it lies a low expanse covering hundreds of acres of lush low plants like pickleweed. This is the Carpinteria salt marsh. One clear summer day, an ecologist named Kevin Lafferty unlocked the fence gate and led me inside. He wanted to show me how a salt marsh works. Lafferty was dressed in a pair of bathing trunks and a fraying T-shirt with fluorescent lion fish on it; he shuffled along the dirt path in flip-flops, with a pair of scuba booties in one hand. I spent a few days all told in the company of Lafferty, and during my entire visit I saw him in nothing more formal. His face was young and his hair was wheat-colored. He has surfed along these beaches since he came to the University of California at Santa Barbara in 1981. It would be hard now to pick him out on a wave as a biology professor instead of a sophomore.

He talked about the marsh as we walked toward the sea on a raised dirt path. "You need some sort of interior space below sea level to get a salt marsh. You can have a river cut a channel and the sea is able to intrude upon it at high tide. That's the standard East Coast version. Or you could have tectonic activity that leads to subsidence." He gestured back inland, up toward the San Ynez Mountains, which loomed over the freeway, fog draped on them like a scarf. "The whole California coast line is a complicated mix of tectonic activity, plus changes in sea level. The basin here is thought to have been flooded by the ocean because it has subsided." The area is now about a foot below sea level, so that the sediments carried by the Santa Monica and Franklin Creeks are dumped in this basin rather than reaching the sea. Each day the high tide pushes its way into the marsh, spilling over the creek banks and flooding this place all the way back to the chain link fence. "If the sea level stayed the same and there was no tectonic activity, this might be dry land in a hundred years. But if the land is continually subsiding, then the sediment can't catch up," says Lafferty. The opposing forces of accumulating sediment, incoming freshwater, and the ebb and flow of sea water have all reached a compromise in the form of this broad, water-logged expanse cut through by channels.

Each day at low tide, the soil bakes in the sun, evaporating its water while holding on to the brine. The soil is actually saltier in places than sea water. In these conditions no trees can survive. Instead, there is a low carpet of tough plants adapted to the salt. Pickleweed, for example, pumps the briny water out of the ground and stocks away the salt in its fruits, using the fresh water left behind. Along the bare mud flats that line the marsh channels, algae grow in dull green varnishes. The algae may look subdued, but they're actually reveling in almost perfect conditions. The mud is packed with nitrogen, phosphorus, and other nutrients carried down from the mountains. Because the bare flats are exposed every time the tide drops, the algae get far more sunlight than they would if they were always submerged. Today at low tide the algae are photosynthesizing merrily. Scattered along the banks are thousands of miniature birthday hats: the conical shells of California horn snails that graze the algae. "They're mowing a fast-growing lawn," Lafferty says.

The many invertebrates here, such as littleneck clams and sand dollars, make good meals for vertebrates. Some fish, like the arrow gobies and the killifish, live in the estuaries year round, huddling in the low water when the tide ebbs and then feeding at high tide, when they're joined by curious stingrays and sharks wandering in from the sea. Today the killifish are the only fish to be seen. They dart around, every now and then turning to one side to expose the brilliant glint of their bellies. Along the banks of the channels are bigger holes, these the size of fists rather than fingers. When the morning sun hits them, crabs slowly crawl out—lined shore crabs, which crack open the snails like walnuts, and fiddler crabs, which slowly raise their giant claws as if saluting the newborn day. There aren't many mammal predators here—the growth of towns like Carpinteria has driven away the mountain lions and bears, leaving only raccoons, weasels, and house cats. But the salt marsh is still a carnival for birds—for Caspian terns, willets, plovers, yellowleg sandpipers, curlews, dowitchers—all picking their way through the feast.

Lafferty looks at all of this, the eating and being eaten, this transmutation of sunlight into different forms of life, and doesn't see it quite the way other ecologists might. A curlew grabs a clam from its hole: "Just got infected," he says. He looks at the bank of snails and says, "More than 40 percent of these snails are infected. They're really just parasites in disguise.

A Precise Horror

There are boxcars of parasite biomass here." He points to the snowy constellation of bird droppings along the bank. "Those are just packages of fluke eggs." He hears the things he's been saying to me and shrugs. "I have a pretty warped perspective."

When Laffterty started graduate school at Santa Barbara in 1986, his perspective wasn't yet warped. If someone had asked him then to figure out the ecology of this salt marsh, he would have studied the things he could see. He would have measured how much algae the snails could eat, he would have added up the number of eggs a female killifish could lay in a year, he would have recorded the number of clams a bird could eat in a day. He would, he now realizes, have completely missed the real drama of this ecosystem because he would have ignored the parasites.

There'd have been nothing unusual in that. For decades, ecologists have waded into bayous, paddled into lakes, and tramped through forests in order to look at two things: the competition for the necessities of life, such as food and water, and the struggle not to be eaten. They surveyed the density of plants and animals, their distribution from young to old, the diversity of species. They drew diagrams of food webs like tangled mobiles. But never did one of those strands lead to a parasite. Ecologists didn't deny that parasites existed, but they thought of them as merely minor hitchhikers. Life could be understood as if it were disease-free. "A lot of ecologists don't like to think about parasites," says Lafferty. "Their vision of the organism stops at the exterior of it."

Few ecologists had bothered to back up their indifference with any data. It didn't matter to them that animals are typically overrun with several different species of parasites. On the other hand, parasitologists had been remiss as well. They had been ogling their parasites in laboratories, but they had no idea what effects they had in the real world.

It turns out that those effects can be huge. Only in the last decade, for instance, have marine biologists discovered that the oceans are swarming with viruses. They had known for a long time that viruses can infect just about any marine life form, from whales to bacteria. But they had thought that there simply weren't many viruses, or that they were too fragile to cause much harm. In fact, viruses are rugged and abundant. Ten billion of them live in the average quart of surface sea water. Their favorite targets are bacteria and phytoplankton, since those are the most abundant hosts in the sea. They also serve as the bottom link in the ocean food chain, devoured by predatory bacteria and protozoa, which are in turn eaten by animals. Now marine biologists realize that this crucial link is very sick. As many as half the bacteria in the ocean are killed by viruses. When a bacterium dies, it bursts apart in a little organic shower. Other bacteria scoop up its remains, in many cases only to be burst open by another virus. A huge amount of the ocean's biomass is stuck in this bacteria-virus-bacteria loop, and it can't feed the rest of the marine food chain. If viruses were to vanish from the sea, it might become crowded with fish and whales.

On land, parasites are just as powerful ecologically. For decades, ecologists who worked on the Serengeti plains thought that the great herds of wildebeest and other grazing mammals there were controlled by two factors: the food that could support them and the predators that kept their population down. Yet, for most of this century it was actually a virus that was most powerful. Known as rinderpest, the disease came to Kenya and Tanzania when infected cattle were imported from the Horn of Africa around 1890. It jumped from the livestock to wildlife and dragged down the population of herbivores, as well as their predators, and kept them down for decades. Only when cattle began to be vaccinated in the 1960s did the mammals of the Serengeti rebound.

Parasites don't even have to kill their hosts to have huge impacts. A parasite may cut down the competitive edge of a species so that it can't drive out a competitor, making it possible for the two species to live side by side. Deer carry a nematode that causes them no harm, but when it gets inside moose, it crawls into their spines and makes them stumble around drunkenly before dying. Without that parasite, the deer wouldn't be able to compete with the moose. And biologists such as Lafferty have shown that the way parasites manipulate their hosts can also have a big effect on the balance of nature.

Going into graduate school, Lafferty thought he had a pretty good idea of the ecology off the California coast, where he had scuba dived since high school (he paid his way through college by scraping mussels off oil rigs). It wasn't until he took a course on parasitology that he had his mind changed. His teacher, Armand Kuris, stunned him by showing how parasites could be found everywhere in the sea. "Here are all these animals I knew and loved as a diver, and when you opened them up they were full of parasites. I realized marine ecology had been missing a big part of the picture."

Lafferty began studying the parasites of the Carpinteria salt marsh. There are many to choose from at Carpinteria—a dozen flukes infect the California horn snail alone—but Lafferty chose the most common one, *Euhaplorchis californiensis*. Birds release *Euhaplorchis* eggs in their droppings, which are eaten by horn snails. The eggs hatch, and the flukes castrate the snail, producing a couple of generations before cercariae come swimming out of their host. The cercariae explore the salt marsh to find their next host, the California killifish. They latch onto its gills and work their way into its fine blood vessels; they crawl deeper into the fish, finding a nerve that they follow until they reach the brain. They don't actually penetrate the killifish's brain but form a thin carpet on top of it, looking like a layer of caviar. There the parasites wait for the fish to be eaten by a shorebird. When they reach its stomach, they then break out of the fish's head and move into the bird's gut, stealing its food from within and sowing eggs in its droppings to be spread into marshes and ponds.

Lafferty wanted to understand what effect this cycle had on the ecology of the salt marsh. Would Carpinteria look the same if there were no flukes? He began his ride around the parasite's cycle at the snail stage. The relationship between fluke and snail is a strange one. It's not a predator-and-prey arrangement. When a lynx kills snowshoe hares, the tender shoots that the dead hares would have eaten are eaten by the survivors, which can use the energy to raise baby hares. But the flukes of Carpinteria don't quite kill their snails. In a genetic sense, the snails are indeed dead, because they can no longer reproduce. But they live on, grazing on algae to feed the flukes inside them. If the snails were truly dead, the algae that they ate would be left for surviving snails to graze on. Instead, the flukes-as-snails are in direct competition with the uninfected snails.

Lafferty set up an experiment to see how the competition played out. "What I'd do is make these cages that had mesh so that water could come in and out, but the snails couldn't go through. The tops were open so the sun could shine through and algae could grow on the bottom. Then I'd bring the snails into the lab and find out who's infected, who's uninfected, and what size they were, and assign the snails to particular cages based on whether they were infected or what size they were. So the cages were all identical except for some factor that was altered. The cages were all located in an area the size of a desk, and that was replicated at eight different sites in the salt marsh."

Lafferty measured how the uninfected snails performed without parasitized snails competing with them. They grew faster, released far more eggs, and could thrive in far more crowded conditions. The results showed Lafferty that in nature, the parasites were competing so intensely that the healthy snails couldn't reproduce fast enough to take full advantage of the salt marsh. In fact, if you were to get rid of the fluke, the snail's overall numbers would nearly double. And this being the real world rather than a lab, that explosion would ripple out through much of the salt marsh ecosystem, thinning out the carpet of algae and making it easier for the predators of snails, such as crabs, to thrive.

After Lafferty earned his Ph.D. in 1991, he continued working with Kuris. He began following the flukes from snails to fish. When Lafferty started working with the parasites, nothing was known about their effects on their killifish hosts. If he scooped up a seine's worth of the fish and dissected them, he found most of them carrying parasites atop their brains. Once they got in, they didn't seem to cause much harm to the fish— the fish didn't even mount an immune response. And as I stood with Lafferty in the salt marsh, looking down at the channels, I certainly couldn't say which killifish were parasitized and which were healthy.

But Lafferty suspected that the flukes might not be passive passengers. Like so many other parasites, they should be taking control of their fates. "Looking at these fish, I didn't notice anything that struck me. But the more I became familiar with all this behavior modification stuff, it seemed like an obvious thing the parasites should be doing," says Lafferty. "They're in a good position to be doing something. Think about a simple molecule like Prozac. It's simple for the flukes to secrete some neurotransmitter."

Lafferty set his student Kimo Morris to establish whether or not the flukes affected the killifish. Lafferty gathered up forty-two fish, brought them into the lab, and dumped them into a seventy-five-gallon aquarium. Morris gazed at the fish for days. He would pick out one and stare at it for half an hour, recording every move it made. When he was done, he'd scoop the fish out and dissect it to see whether its brain was caked with parasites or not. And then he'd meditate on another killifish.

What was hidden to the naked eye came leaping out of the data. As killifish search for prey they alternate between hovering and darting around. But every now and then, Morris would spot a fish shimmying, jerking, flashing its belly as it swam on one side, or darting close to the surface. These might be risky things for a fish to do if a bird was scanning the water. And Morris's vigil had revealed that fish with parasites inside them were four times more likely to shimmy, jerk, flash, and surface than their healthy counterparts. Since then, Lafferty has been working with a molecular biologist to figure out how the parasites make their hosts dance. They've found that the flukes can pump out powerful molecular signals, known as fibroblast growth factors, which can interfere with the growth of nerves. They could turn out to be the parasite's Prozac.

Lafferty decided to see what effect this manipulation had on the salt marsh ecology. "Once we saw that the behavior was different, it was obvious that the field experiments had to follow," he says. Lafferty wanted to see if what Morris might perceive as an unusual behavior could really translate into a better chance that a fish would be eaten by a bird—and not a bird stuck in a lab cage but one free to fly to another marsh if it was so inclined. He and Morris set up a series of pens that were both open to the sky and flush on one side with the shore, so that fish couldn't escape, but birds could easily land in the pens or simply wade into them. They filled both pens with a mix of infected shimmying fish and healthy ones, and covered one with netting to protect it from birds.

For two days they watched the pens, not knowing whether birds would even bother with them. Then a great egret waded into the open pen, stepping slowly, as if in deep thought. It stared into the muddy water and then struck a few times, the last time bringing up a killifish.

After three weeks, Lafferty and Morris gathered the fish out of the pens. They brought them back into the lab to look inside their skulls. The results were even more stark than Morris's fish-watching: the birds were not four times more likely to select one of the flailing, parasitized fish, but thirty times. Either their eye is far keener than Morris's, or perhaps they are that much lazier.

A Precise Horror

But why would birds pick so many sick fish when they were virtually guaranteeing themselves an intestinal parasite? The flukes do take a toll on the birds, but a relatively small one. It's in the parasite's interest, after all, for the bird to be healthy enough to fly, so that it can carry the fluke to other salt marshes that it can colonize. If the bird scrupulously avoided infected killifish, it might stay healthy, but it would also go hungry. The parasites make so much food available to it that their benefits far outweigh their costs.

Armand Kuris was stunned by what his former student had found. "What blew me away was the conservative estimate that they increased the susceptibility to predation by thirty times. *Thirty times.* So now I step back, and I look at the birds flitting around out there and think: Could we have those birds out there if it were thirty times harder for them to get their food? It was that that made me go from thinking that behavior modification was just a great story to thinking that it's really powerful—it may be running a large part of the waterbird ecology. Is there anything to birds other than this?"

This sort of power isn't limited to a salt marsh on the California coast. Two thousand miles away from the Carpinteria salt marshes, ecologist Greta Aeby has been scuba diving along Hawaii's coral reefs. Corals are actually colonies of animals, each a soft polyp lodged in a hard chalky scaffolding. The polyp can reach out into the seawater to filter out food or to spawn, but then it retracts back into the safety of its armor. A marine fluke called *Podocotyloides stenometra* begins its life inside clams that live around the reef; then it invades coral polyps for the next stage of its cycle. From there it needs to get into the intestines of the butterfly fish, which graze the corals. Butterfly fish have to put a lot of effort into nibbling at what little flesh of the polyps is exposed above their drab brown exoskeleton.

A parasite can't make coral dance like killifish in order to get the attention of its next host. But Aeby has found that

Podocotyloides manages to make some changes to the polyp that are just as effective. When the fluke gets inside the coral, the polyp swells up and changes from its normal brown to a bright pink. At the same time it grows a network of calcium carbonate spikes that keep it from retracting. As a result, the swollen brilliant polyp dangles out, making it an easy pick for a passing butterfly fish. In fact, when Aeby put butterfly fish into a tank with healthy and parasitized corals, 80 percent of their bites were directed to the sick coral. In half an hour one fish can swallow 340 flukes.

But Aeby has found that the alliances in her ecosystem are different from the ones that Lafferty has uncovered in salt marshes. When a killifish brings a fluke to a bird, the killifish dies in the process. But corals consist of colonies of clones and when an individual polyp infested with a fluke dies, it is replaced by a healthy new one. An infected polyp can't feed or reproduce, so allowing a fluke to fester inside it is a drain on the colony, slowing its growth. If a butterfly fish prunes the coral, it can perform as well as a healthy coral. It's to the coral's advantage to get rid of its sick polyps, which may mean that the coral is actually contributing to the color or spikes in order to make it easier for the butterfly fish to spot. Lafferty found a case in which a parasite and its final bird host were allied; here, Aeby has found a case where the intermediate host and the parasite work together.

Discovering parasites at work in ecosystems can feel a bit like watching in terror as a bank robbery unfolds and then looking across the street and seeing a movie crew with its cameras and boom mikes. Birds are being guided to their meals, and fish are choosing their coral polyps, thanks to the advertisement of flukes. Uncovering these effects is hard work, and only a few examples have been documented. But they're enough to suggest that parasites can cast some of the hoariest notions of ecology into doubt. We tend to think of predators as keeping a herd of prey healthy by weeding out the slowest ones. That's not what's happening in Lafferty's salt marsh, or even among those icons of predator and prey, the wolf and the moose.

Wolves are the final hosts for one of the smallest tapeworms in the world, *Echinococcus granulosus.* Far from a ticker-tape ribbon, it's lucky if it gets to be a quarter of an inch long as an adult. It doesn't cause its final host much harm, but its eggs can be vastly vicious. They are eaten by herbivores such as moose, where they slowly transform themselves into cysts in which thirty individuals may sit. They will keep growing if there's no bone in their way. When they accidentally end up in humans, they have been known to grow so big that they've contained fifteen quarts of fluid and millions of baby tapeworms.

One of the tapeworm's favorite sites for forming its cyst is the lungs. A moose may carry several in its lungs, each tearing through its bronchial tubes and blood vessels. As a result, when wolves sweep down on a herd of moose, they're more likely to pick out the slow, wheezing one and kill it. It's even possible that these moose tapeworms can create the same kind of scent used by rat tapeworms to lure beetles. Instead of leaving the scent in droppings, though, the moose tapeworms could release their aroma with their host's every breath. In any case, the result is that the tapeworm brings the wolf to the moose so that it can get into the wolf. The thinning of the herd is an illusion, not the service of the predator but the side effect of a tapeworm traveling through its life.

•   •   •

On my way to see Lafferty, I stopped one night in a hotel in Riverside, California. It had originally been a Spanish mission, and after unpacking, I prowled around the old shrines, explored the hidden passageways surrounded by vines and palms, crossed the hushed stone courtyard. I came back to my room feeling utterly alone. I turned on the television for company. An episode of *The X-Files* was on. As well as I could figure out, an FBI man had suddenly turned gloomy and wouldn't return anyone's phone calls. When another agent tracked him down and confronted him, the gloomy man threw him to the floor and brought his face close to his, opening his mouth. With wonderful creaking and slithering noises, a scorpionish creature crawled out of his throat and climbed into the other agent's mouth.

I didn't feel so lonely after that. Some television screenwriter had parasites on his mind as well. It occurred to me that parasites were the basis for a lot of science fiction novels, of movies and television shows. And I was struck by the fact that these parasites were dangerous because they could manipulate their hosts, just as parasites can in reality. When I got back home I started renting videos. I told my friends, and they'd tell me about other movies I should see, books to read. It got to be a gruesome marathon. The oldest entry I could find was Robert Heinlein's *The Puppetmasters*, a 1955 novel. A spaceship full of aliens travels from Saturn's moon Titan and lands near Kansas City. But the aliens inside aren't the standard-issue 1950s hairless bipeds; they're pulsating jellyfish-like creatures that latch onto people's spines. Hiding underneath the clothes of their hosts, they tap into their brains and force them to help spread the parasites across the planet. The fight against them is a bit ludicrous, with the government forcing everyone to walk around practically naked to be sure they're not carrying an alien. Humanity is saved when the army finally finds a virus that can kill the parasites, and the book closes with a fleet of spaceships leaving Earth for Titan to exterminate the parasites for good. It's a stiff, peculiar book—the only one I've read that ends with the battle cry "Death and Destruction!"

The Puppetmasters was turned into a pretty mediocre movie in 1994, but its essence—the notion of humans harboring giant parasites—has become a Hollywood institution. Parasites are a part of our shared dramatic language, just as they were in Greek comedies. Any blockbuster can rest its plot on parasites without anyone's worrying that it will seem too esoteric. One of the biggest movies of 1998, *The Faculty*, takes place in a high school where parasites from another planet are taking over the bodies and minds of teachers and students. These fluke-like things sprout tentacles and tendrils, and they pull themselves into their new hosts through their mouths or ears. Their hosts change from frazzled teachers and sulking, violent kids to glazed-eyed upstanding citizens who try to spread the parasite to new hosts. It's up to the assorted losers of the school—drug dealers, geeks, and dropouts—to save the world from the invasion.

Parasites got their first big break at the movies almost twenty years earlier, in the 1979 movie *Alien*. A spaceship hauling ore stops off to investigate a crash on a lifeless planet. The crew discovers an alien ship that has been destroyed in a ruthless attack, and nearby they come across a clutch of eggs. One of the crew, a man named Kane, takes a close look at one of the eggs, and a giant crablike thing bursts out of it, clamping to his face and wrapping a tail around his neck. His crewmates bring him back to their ship, alive but comatose. When the ship's doctor tries to get the thing off him, it tightens its tail around Kane's neck. The next day it has disappeared, and Kane seems fine. He gets up and eats voraciously, to all appearances normal. Of course, no movie monster ever just disappears. This one has been devouring Kane's guts, and before long he suddenly clutches his stomach, writhing and screaming, and a little knobby-headed alien pierces through his skin and leaps out. As the parasitic wasp is to the caterpillar, so this alien is to humans.

Alien may have made Hollywood safe for parasites, but a lot of the conceptual legwork had already been accomplished four years earlier in a low-budget, little-seen movie directed by David Cronenberg called *Shivers*. It is set on Starlight Island, an immaculate high-rise building on an island outside Montreal. "Sail through life in quiet and comfort," says the soothing voice-over on a commercial for the building. But the isolated quiet and comfort is destroyed by an engineered parasite. It's the work of one Dr. Hobbs. Dr. Hobbs originally set out to create parasites that could play the role of organ transplants. A parasite could be connected to a person's circulatory system and filter blood like a kidney, for example, while taking only a little blood to keep itself alive. But Dr. Hobbs also has a secret agenda: he's decided that man is an animal that thinks too much, and he wants to turn the world into one giant orgy. To that end he fashions a creature that will be a combined aphrodisiac and venereal disease: a parasite that will make its hosts sexually voracious and will be spread during sex.

He implants it in a young woman he has been having an affair with, a woman who lives on Starlight Island. She sleeps with some of the other men in the building and spreads the parasite. A stubby worm the size of a child's foot, it lives in people's guts and passes from mouth to mouth during a kiss. It transforms people into sexual monsters, attacking each other in apartments, laundry rooms, elevators. Rape, incest, and all sorts of other depravity erupt.

The physician for Starlight Island spends most of the movie trying to stop the parasite from spreading. At one point he has to shoot a man attacking his nurse (and girlfriend), and they escape to the basement. As they cower there, the nurse tells him that she had a dream the night before in which she was making love to an old man. The old man told her that everything is erotic, everything is sexual, "that disease is love of two alien kinds of creatures for each other." Whereupon she tries to kiss the doctor, with a parasite crouched in her mouth ready to spring. He knocks her out cold. He tries to escape the building, but hordes of infected hosts ring him in and herd him into the building's swimming pool. His nurse is there, and she finally gives him a fatal kiss. Later that night, all the residents drive out of the garage and leave the island, to spread the parasite and its mayhem throughout the city.

As I watched these movies, I was struck by how easy it was to translate biological reality into movie horror. The creature in *Alien* comes as no surprise to the entomologist who studies parasitic wasps. Heinlein may not have known that parasites can take over the behavior of their hosts, but he nailed the essence of their control. It may seem ridiculous that the parasites in *Shivers* can spread themselves by making people have sex, but it's no more ridiculous than what actual parasites do. The fungus that I discussed earlier, which infects flies and forces them to climb up grass in the evening, actually uses a second trick to spread itself as well. It makes the corpse of its host a sexual magnet. Something about the fly—something brought about by the fungus itself—makes it irresistible to uninfected male flies. They will try to mate with it, preferring it to living flies. As they grope the corpse they become covered with spores themselves. When they die, they themselves become irresistible. When will someone make their movie?

Of course, these parasites are more than just parasites. In *Shivers*, Cronenberg uses them to expose the sexual tension buried under the blandness of modern life. In *The Faculty*, parasites represent the stupefying conformity of high school, which only outsiders can fight. And in *The Puppetmasters*, written in the McCarthyite fifties, the parasites are Communism: they hide within ordinary-seeming people, they spread silently across the United States, and they have to be destroyed by any means necessary. At one point the narrator says, "I wonder why the titans [the narrator's name for the aliens] had not attacked Russia first; Stalinism seemed tailormade for them. On second thought, I wondered if they had. On third thought, I wondered what difference it would make; the people behind the Curtain had had their minds enslaved and parasites riding them for three generations."

But all these works do have something in common: they play on a universal, deep-seated fear of parasites. This horror is new, and for that reason it's interesting. There was a time when parasites were treated with contempt, when they stood for the undesirable, weak elements of society that got in the way of its progress. Now the parasites have gone from weak to strong, and now fear has replaced contempt. Psychiatrists actually recognize a condition they call delusional parasitosis—a terror of being attacked by parasites. The old parasite metaphors, the ones used by people like Hitler and Drummond, were remarkably precise in their biology. And, judging from movies like *Alien* and *The Faculty*, so is the new one. It is not just a fear of being killed; it's a fear of being controlled from within by something other than our own minds, being used for something else's ends. It's a fear of becoming a flour beetle controlled by a tapeworm.

This precise horror of parasites has its roots in how we now see our relationship to the natural world. Before the nineteenth century, Western thought saw humans as distinct from the rest of life, created by God with a divine soul in the first week of Genesis. It became harder to keep that dividing line fixed as scientists compared our bodies with those of apes and found the differences to be pretty minor. And then Darwin explained why: humans and apes are related by common descent, as is all of life. The twentieth century has given his realization a fine-grained detail, moving from bones and organs down to cells and proteins. Our DNA is only a shade different from that of chimpanzees. And like a chimpanzee, or a turtle or a lamprey, we have brains that consist of crackling neurons and flowing neurotransmitters. These discoveries may give some comfort if you look at them one way: we belong on this planet as much as the oak and the coral reef, and we should learn to get along better with the rest of the family of life.

But look at them another way, and they bring horror. Copernicus took the Earth out of the center of the universe, and now we have to accept the fact that we live on a watery grain in an overwhelming void. Biologists like Darwin did a similar thing, taking humanity out of its privileged place in the living world—a biological Copernicanism. We still go through life pretending that we are exalted above other animals, but we know that we too are collections of cells that work together,

A Precise Horror

kept harmonized not by an angel but by chemical signals. If an organism can control those signals—an organism like a parasite—then it can control us. Parasites look at us coldly—as food, or perhaps as a vehicle. When an alien bursts out of a movie actor's chest, it bursts through our pretenses to be more than brilliant creatures. It is nature itself that is bursting through, and it terrifies us.

5

The Great Step Inward

Whence, thinkest thou, kings and parasites arose?
—Percy Bysshe Shelley, *Queen Mab*

There are billion-year-old secrets at the University of Pennsylvania, but they are well hidden from view in the laboratory of a biologist named David Roos. The sunlight of a soft Philadelphia sky flows through high windows into the lab, where Roos's graduate students are laying flasks of cherry-colored liquids under microscopes, kneading data on computers, clicking pipettes in test tubes, and working in incubator rooms, cool rooms, warm rooms. Overhead, the sunlight strikes the vines and aloe plants on the shelves. The plants take in the summer light, each photon falling onto the surface of a microscopic, blob-shaped structure called a chloroplast. A chloroplast is essentially a solar-powered factory. It uses the energy of the light to manufacture new molecules out of raw materials such as carbon dioxide and water. The new molecules are trundled out of the chloroplasts and used by the plants to sprout new roots, to send out new feelers along the shelf. Below them, Roos's students work furiously, discovering the hidden biochemistry of a parasite and publishing scientific papers, as if within them the sun were also driving some kind of intellectual photosynthesis. At a time like this, in a place like this, who has time to think about ancient history?

David Roos runs the lab from an office lodged at its center. He's a young man with a curly mat of black hair and a chipped front tooth. He speaks coolly, comfortingly, his answers rolling out in paragraphs and pages with references ahead and back from the subject at hand, with hardly a pause for collecting thoughts. On the sunny day I visited, he was explaining to me how he came to study the parasite that he carries by the thousands in his own brain: *Toxoplasma gondii*. Overhead are charcoal drawings of human figures, a reminder of Roos's days as an art student in college. That came after a stint after high school as a computer programmer—"I thought I wouldn't go to college, since I was having so much fun and making so much money as a programmer, but that got old fairly quickly"—and before Roos took up biology. When he began studying biology, he contemplated working on parasites. "There's no more interesting question biologically than how does one organism survive off of another, especially inside another cell? But as a graduate student I looked around and talked to a couple of labs, and the systems just seemed so archaic."

By this, Roos meant that parasitologists had a harder time with husbandry than other biologists. A lot of scientists who study how animals develop from fertilized eggs, for example, study the fruit fly. If they find an interesting mutation in a fly, they know how to breed a line of them that all carry the same mutation; they have the tools to isolate the mutated gene, to shut that gene down or replace it with a different version. With these tools, biologists can map out the web of interactions that turn a single cell into a noble insect. But parasitologists struggle just to keep parasites alive in a lab, and breeding interesting strains is often impossible. Fruit fly biologists have a giant toolbox at their disposal. Parasitologists have been stuck with a broken hammer and a toothless saw.

The frustration didn't appeal to Roos, so he went off to work in graduate school on viruses, and later on mammalian cells. His work paid off well, landing him a job at Penn, but by then he wanted something new to study. He learned that in the years he had stayed away from parasites, other researchers had had some early success in using them like fruit flies. One parasite looked particularly promising: *Toxoplasma*. It might not have the cachet of its close relative *Plasmodium*—the parasite that causes malaria, a sophisticated creature that can turn a barren red blood cell into a home in a matter of hours—but it seemed to take well to life in the lab. Perhaps it could act as a model for malaria, since many of their proteins worked in similar ways. "I thought, maybe very naively, that one of the reasons people had not worked on *Toxoplasma* in the past was that it was rather boring," Roos said. "Like anybody else, biologists like to work on sexy topics. But maybe if this organism is so boring—meaning more or less like things we're more familiar with—it wouldn't require completely reinventing the wheel to develop genetic tools."

Roos started building the tools, and he found success unnervingly simple. "Some people think we have golden hands in my lab, but in truth we work on an easy organism," he says. His lab learned how to riddle the parasite with mutations, how to switch one gene with a new one, how to see the parasite more clearly than before. Within a few years they were able to start using their tools to ask questions, such as exactly how *Toxoplasma* invades cells, or why some drugs kill *Toxoplasma* and *Plasmodium*, while the parasites manage to resist others.

In 1993, Roos began studying a drug that kills both parasites, called clindamycin. It's not used to cure malaria, though, because it takes too long to kill *Plasmodium*; instead, it's chiefly used against *Toxoplasma* in AIDS victims who need a drug they can take for years without side effects. "The funny thing about clindamycin," Roos says, "is that it shouldn't work."

Clindamycin is actually used mostly as an antibiotic to kill bacteria, which it does by clogging up the bacteria's protein-building structures, known as ribosomes. "Eukaryote cells have quite different ribosomes, and clindamycin doesn't interfere with them, which is good, because otherwise it would kill you. That's what makes it a good drug. Now *Toxoplasma*, these guys aren't bacteria. They have a nucleus, they have mitchondria." (Mitochondria are compartments where eukaryote cells generate their energy.) "They're clearly more closely related to you and me than to bacteria."

And yet, clindamycin kills *Toxoplasma*, and *Plasmodium* as well. How it killed them no one knew. Scientists knew that they didn't affect the regular ribosomes in the parasites. But eukaryotes also carry a few extra ribosomes in their mitochondria that are different from the rest. Mitochondria carry their own DNA, which they use to build their own ribosomes, among other things. Yet, researchers found that clindamycin left the ribosomes of mitochondria unharmed as well.

Roos rememberd that *Toxoplasma* actually had a third set of DNA. In the 1970s, scientists had discovered a circle of genes that didn't belong to its nucleus or its mitochondria. This orphan DNA contained the recipe for a third ribosome. Perhaps, Roos thought, clindamycin attacked the third ribosome and killed the parasites in the process. He and his students destroyed the circle of DNA and discovered that indeed *Toxoplasma* couldn't survive without it.

But what exactly was this ring of genes? Roos and his students discovered that it sat inside a structure floating close by the parasite's nucleus. In the past, scientists had given the structure many names—the Spherical Body, the Golgi Adjunct, the Multi-membraned Body—all of which may make you think they knew what it was for. They didn't.

Roos now knew it was for housing the genes that make *Toxoplasma* vulnerable to clindamycin. But he didn't know yet what the ribosome that the genes made was for. To get some insight, he compared the genes to other genes in *Toxoplasma* and other microbes. The closest match he found was not among the genes inside *Toxoplasma*'s nucleus or mitochondria. It was the chloroplasts in plants, those solar-powered factories that make the plants on the laboratory shelves grow. "They look for all the world like a green plant," says Roos.

Roos had hoped to figure out how *Toxoplasma* and *Plasmodium* die like bacteria, even though they live like us. Now he had simply traded one puzzle for another: How can malaria be a cousin to ivy?

•　　•　　•

To nineteenth-century biologists such as Lankester, parasites got to be the way they are now by degeneration. Their evolutions were tales of loss, of the abandonment of all the adaptations that made an energetic, free-living existence possible, of settling for a spoon-fed dinner. In this century, that notion of degeneration has hung on; for decades, evolutionary biologists simply thought that the story of parasite evolution was not worth thinking about compared with sagas like the origin of flight or the enfolding of the brain. Yet, the ability of *Trichinella* to make its host build itself a nursery in its muscles, of *Sacculina* to make a male crab into its mother, of blood flukes to become blood-invisible—all of these are adapations produced by evolution. Many parasitologists don't have evolution as their main business; they study parasites as they live today. And yet, evolution elbows its way into their work.

Such is the case with David Roos: the only way he can understand what *Toxoplasma* is today, and how it is that malaria is a green disease, is to plunge back hundreds of millions of years. These sorts of histories are just as fascinating as those of free-living animals. They are tangled up with the evolution of the rest of life, going back 4 billion years. In fact, the history of parasites is, to a great extent, the history of life itself.

Reconstructing that history isn't easy. Parasites tend to be squishy or crunchy—two conditions that don't augur well for fossils. Every few million years, a parasitic wasp may stumble into a blob of amber, or a male crab feminized by a parasitic barnacle may leave behind its transgendered fossil, but for the most part parasites vanish in the rotting tissues of their hosts. Rocks don't have a monopoly on clues to life's history, though. Evolution has formed a vast tree, and biologists today can inspect its leafy tips. By comparing the biological features they find there, they can work their way back to the crooks of branches, to the tree's base.

Biologists draw the branches of this tree by figuring out which species are most closely related to one another. Their close heritage shows that they must have diverged from a common ancestor more recently than from other species. To see this kinship, biologists look at the similarities and differences among organisms, judging which ones are the result of common descent or the illusions of evolution. A duck, an eagle, and a bat all have wings, but the duck and the eagle are much more closely related. The evidence is in their wings: on birds they consist of feathers hanging from a fused hand; a bat has membranes stretched over long fingers. The fact that bats are hairy, give birth to live young, and nurse them with milk helps show that despite their wings, they're actually more closely related to us and other mammals than to a bird.

Flesh and bone can say only so much, though. They do not say definitively whether bats are closer cousins to primates or to tree shrews, for instance. And for organisms that don't have flesh or bone, they say nothing at all. That silence has pushed biologists in the past twenty-five years to compare the protein and DNA of organisms rather than wings or antlers. They have learned how to sequence the genes and compare them with the help of computers. This approach brings its own pitfalls—

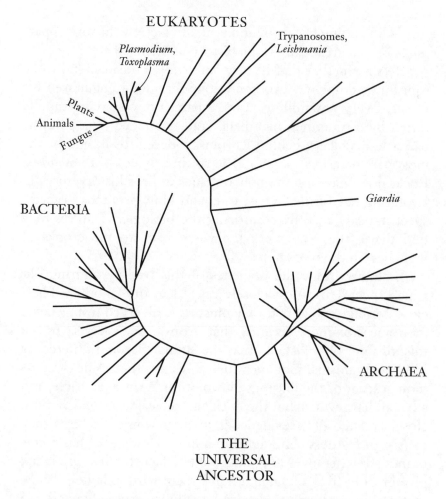

The tree of life, showing the evolutionary position
of a few parasites (adapted with permission from Pace, 1998)

genes can sometimes create trees as confusing as flesh and
bone—but while they may be provisional, they have allowed biologists to look for the first time with one grand sweep of the
eye at all of life.

The base of the tree represents the origin of life. Many of
the organisms that occupy the branches closest to the base live

today in scalding water, often around hydrothermal vents. That suggests that life may have gotten its start in such a place 4 billion years ago. Gene-like molecules may have assembled inside little fatty capsules or perhaps in oily films coating the sides of the vents. After untold millions of years, the first true organisms formed, bacteria-like things that carried genes floating loose inside their walls. Out of these bacterial beginnings, life began to diverge into separate lineages. The Archaea continued a basically bacteria-like kind of life, while a third branch—the eukaryotes with their DNA balled up tight in a nucleus and their power coming from mitochondria—took on a drastically different form.

Parasites, according to the traditional definition of the word (the creatures that cause malaria and sleeping sickness, that cram into guts and livers, that burst out of caterpillars as if their hosts were giant birthday cakes), all sit on branches on the eukaryote part of the tree. They have abandoned a life in the sea or on land for one inside other eukaryotes. They include organisms separated by vast evolutionary gulfs from ourselves—trypanosomes and *Giardia* branched off on their own separate destinies at the dawn of the age of eukaryotes, over two billion years ago. Among the parasites there are also much closer relatives, such as fungi and plants. Parasitic animals, such as blood flukes and wasps, are practically our kissing cousins. Parasitism is scattered across the eukaryote domain, a way of life that lineages have independently adopted and have found to be immensely profitable for many hundreds of millions of years.

Yet, this tree also makes it clear just how shallow the conventional definition of *parasite* is. Why should the name be restricted to organisms that are found on one of the three great branches of life? Nineteenth-century biologists were right to call infectious bacteria parasites. Just as some eukaryotes abandoned the free-living life, so did certain bacteria such as *Salmonella* and *Escherichia coli*, while other bacteria have kept up their independence in oceans, swamps, and deserts—even under Antarctic ice. The difference is only in genealogy, not lifestyle.

And even this definition of parasites is too parochial. Nowhere on this tree, for instance, can you find a flu virus. That's because viruses aren't, strictly speaking, living things. They have no inner metabolism and can't reproduce on their own. They are nothing more than protein shells, which carry in them the equipment necessary to get into cells and then use the cell's own machinery to make copies of themselves. Yet, viruses have the same sorts of parasitic hallmarks you could find in creatures like blood flukes—they thrive at their host's expense, they use some of the same tricks to evade the immune system, and they can sometimes even change their hosts' behavior to increase their spread.

In the 1970s, the English biologist Richard Dawkins made viruses less of a paradox. Viruses may not be alive in the traditional sense, but they get the basic job of life done: they replicate their genes. Animals and microbes exist, Dawkins argued, to do the same thing. We should think of their bodies, their metabolism, their behavior all as vehicles that genes build in order to get themselves replicated. In that sense, a human brain is no different from the protein coat that allows a virus to slip inside a cell. This view of life is a controversial one, and many biologists believe it downplays the importance of life's complexity. But it works very well when it comes to parasitism. For Dawkins, parasitism is not what some particular flea or thorny-headed worm does. Parasitism is any arrangement in which one set of DNA is replicated with the help of—and at the expense of—another set of DNA.

That DNA can even be part of your own genes. Huge swaths of human genetic material do nothing for the good of the body they're in. They don't make hair, they don't make hemoglobin, they don't even help other genes do their job. They consist of little more than the instructions for getting themselves replicated faster than the rest of the genome. Some of them produce enzymes that slice them free and then insert them at another point in your genes. Soon the gap they leave behind is visited by proteins that search for damaged DNA. Because human genes come in pairs, these proteins can use the undamaged copy as a guide, and rebuild the stretch that disappeared. In the end, there are two copies of the jumping DNA.

These chunks of wandering genetic material are sometimes called selfish DNA or genetic parasites. They use their host—their fellow genes—to get themselves replicated. Like more conventional parasites, genetic parasites can harm their host. As they insert themselves at random places in the genome, they can cause diseases. Because genetic parasites can replicate at a faster rate than their fellow genes, they have swamped the genome of many hosts, including humans.

Parents pass their genetic parasites down to their children, and it's possible therefore to sort selfish DNA into families, descendants of common ancestors that lived within the common ancestors of their hosts. Genetic parasites have their own dynasties that rise and fall. When a founder first turns up in a new host's DNA, it starts copying itself at an explosive pace, packing its host gene with parasites. (I speak here of an explosion over evolutionary time—perhaps thousands of years.) Genetic parasites are sloppy duplicators, though, and they often make defective copies of themselves. These misfits can't replicate themselves and simply clog up their host's DNA. Genetic parasites are thus always risking self-inflicted extinction.

They can escape this dead end with little bursts of evolutionary renewal. Some of them steal genes from their host that allow them to build protein shells. They become viruses that can break free of their own cell and infect other ones. Some of these breakaways can even infect new species. They probably get carried away by parasites (such as mites) that take them to their new host, although some of the jumps are so long that it's hard to know how they could possibly happen. How is it, for instance, that a freshwater flatworm has the same genetic parasites as a hydra living in the ocean, and a beetle living on land?

Viruses and genetic parasites may be common today, but 4 billion years ago parasitism might have been even more rampant. A typical organism alive today, be it a bacterium or a redwood, carries genes that are organized into powerful coalitions. They can copy themselves accurately into a new generation, and they can put up a fight against cheating genes. But when the Earth was young, some biologists think that genes were barely organized and couldn't cooperate very well. Genes moved fluidly from one microbe to the next, sliding in and out of genomes through a sort of global microbial network. Any genes that could trick others into replicating them would be rewarded by natural selection and spread. Eventually the coalitions of genes got organized into separate organisms, but they were still trading DNA around so promiscuously that a biologist would have a hard time classifying them into separate species.

In spite of the assaults, true organisms did manage to evolve. Probably their genes evolved to a point where they all worked together well and could shut out cheating genes, and they could faithfully replicate themselves. It was probably at this time that life began to diverge into three great branches: bacteria, Archaea, and eukaryotes. Some of those early microbes found their energy in the chemicals growing along hydrothermal vents. As hundreds of millions of years drifted by, some lineages of bacteria became able to capture the energy of light. Other bacteria scavenged their microbial dung. Others evolved into killers, swallowing up the self-sufficient bacteria. Genetic parasites still lived off these different kinds of microbes, although their hosts had begun to get the upper hand.

But with every level of complexity that life achieved, a new kind of parasite emerged. When true organisms evolved, some of them became parasites. There are a few plausible stories of how they first evolved, and they may all turn out to be true in one case or another. One story begins with microbial predators swallowing what should have been their next meal. They opened up a cavity in their membrane and engulfed their prey;

they prepared to carve it up, but for some reason, that was as far as their meals got. The prey sat in the predator's microbial belly, indigestible.

Now the tables were turned—the prey turned out to be able to get a little nutrition from its failed predator before it was spat out. That extra food, that brief shelter from more successful predators, helped the prey reproduce more quickly than it would have otherwise. Natural selection would make the genes that helped it survive inside the predator became more common. They were joined by other genes that helped the prey actually seek out its predator, to open those cavities in the predator's membrane by themselves. The prey spent more and more time inside the predator and gradually abandoned its free-living ways. Now it became the predators that had to fight off the prey, putting more and more effort into expelling them. If the cost of trying to fight off the invasion of parasites became too great, it would have benefited some hosts to make their parasites full-time guests. When the host divided, the parasite copied its own DNA and passed it down through the generations.

Once brought together this way, parasite and host can take their relationship in any one of several directions. The parasite may go on making its host's life miserable, or it may instead become useful to the host, perhaps secreting some protein that the host can use. After many generations together, the lines between parasite and host may begin to blur. Some of the DNA of the parasite is accidentally ferried into the host's own genes. The parasite itself may shrivel away to a few essential functions. The two organisms become essentially one.

Darwin never imagined this sort of fusion of life. He thought of life as an ever-branching tree, something like the tree shown on page 124. But biologists now recognize that they need to braid some of the branches together.

Scientists are now sequencing the full battery of genes in many microbes, and in them they can see signs of the choices that parasites have taken. Among the fully sequenced species is

Rickettsia prowazekii, a bacterium that causes typhus. It invades cells, soaks up their nutrients and consumes their oxygen, multiplies like mad, and bursts its hosts open. Its DNA looks remarkably like the DNA in mitochondria, the organelles that provide every cell in our body with energy. A primordial free-living bacteria must have been the ancestor of both *Rickettsia* and mitochondria perhaps 3 billion years ago. Some of its descendants ended up passing through the earliest eukaryotes. The branch that led to *Rickettsia* evolved down the vicious path, while mitochondria's ancestors eventually settled peacefully inside their hosts. Mitochondria was a fortunate parasite for our ancestors to gain. Photosynthesizing bacteria were gradually filling the atmosphere with oxygen, and mitochondria let eukaryotes breathe it.

Today's eukaryotes are the product of a slow orgy of feasting and infection. After mitochondria invaded, several branches of eukaryotes all gained more bacteria of their own. These bacteria were photosynthetic, and their hosts stripped them down to their bare sun-harnessing essence, the chloroplast. These eukaryotes gave rise to algae and land plants, which added even more oxygen to the air. We can breathe oxygen, and plants can produce it in vast quantities, thanks to the parasites inside our cells.

This billion-year-old drama explains how malaria came to be a green disease. Some ancient eukaryote swallowed a photosynthesizing bacteria and became a sunlight-gathering alga. Millions of years later one of these algae was devoured by a second eukaryote. This new host gutted the alga, casting away its nucleus and its mitochondria, keeping only the chloroplast. That thief of a thief was the ancestor of *Plasmodium* and *Toxoplasma*. And this Russian-doll sequence of events explains why you can cure malaria with an antibiotic that kills bacteria: because *Plasmodium* has a former bacterium inside it doing some vital business.

It's hard to know what exactly that ancient parasite did with its newfound chloroplasts. Perhaps it used them to live like a plant by photosynthesis. But that's not the only possibility, because chloroplasts in plants do more than harness sunlight. They make many compounds, including fatty acids (the sort of molecules that constitute olive oil, for example). David Roos and his colleagues have speculated that in *Plasmodium* and *Toxoplasma*, their remnant of a chloroplast still makes these fatty acids and that the parasites use them to enshroud themselves inside their host cells. Clindamycin may be lethal to the parasite because it destroys *Plasmodium*'s bubble.

One thing is clear, though: that ancestor of *Plasmodium* and *Toxoplasma* didn't live inside animals. A billion years ago, there weren't any animals yet to parasitize. At the time, single-celled creatures were only just beginning combining into colonies and collectives. Many of the first multicellular creatures were like nothing alive today. Some of them looked like inflatable mattresses or the ornate coins of some ancient kingdom. It wasn't until about 700 million years ago that the first kinds of animals we see today arose: corals, jellyfishes, arthropods. Meanwhile, algae began organizing into more complicated forms, giving rise to plants, and about 500 million years ago they moved on shore, forming a mossy carpet and later evolving into low-stalked plants, and finally trees. Soon afterward, animals came on shore as well—centipedes and insects and other invertebrates by 450 million years ago, and the first lumbering vertebrates 360 million years ago.

Multicellular organisms created a seductive new world for parasites to explore. They concentrated food into big, dense bodies that were stable homes for weeks or years at a time. The animals of the Cambrian oceans attracted protozoa like *Plasmodium* as well as bacteria and viruses and fungi. And once again, a new kind of parasite came into existence: animals themselves evolved to live inside other animals. Flatworms made their way into crustaceans, where they diversified into flukes, tapeworms, and other parasites. Crabs, insects, arachnids—at least fifty times other lineages of animals followed suit.

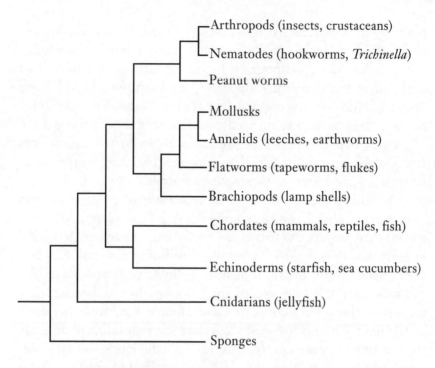

The evolutionary relationships of animals
(adapted with permission from Knoll and Carroll, 1999)

The parasites evolved quickly within their hosts into forms quite unlike their ancestors. Relatives of jellyfish began to parasitize fish, and stripped themselves down into little sporelike shapes, which today plague the trout of American rivers with whirling disease. As their hosts became bigger and more widespread—growing to towering trees, ant colonies millions strong, marine reptiles eighty feet long—parasites enjoyed an ever-expanding habitat. After the first flush of success at the dawn of life, after the brutal clamp-down as hosts became better organized, now came a new golden age for parasites.

Our own lineage, the vertebrates, hasn't done a very good job at becoming parasites. Among the few that have are some species of catfish in the rivers of Latin America. The most famous one of them is the candiru, a pencil-thin fish. It earns its fame by attacking people who urinate in rivers. It follows the odor of their urine and rams itself into their urethra. Once it sinks its teeth into a penis or a vagina, it's almost impossible to get out. Attacking people is not how the candiru makes a living, though; it usually feeds on other fish, working its way under their gill flaps and sucking blood from the delicate vessels underneath. After a few minutes it drops off and looks for another fish to make its host. Other species have an even more parasitic way of life. When fish are caught in Latin America, they're sometimes found with inch-long catfish lodged in their gills. Those little fish may spend most of their lives there, feeding on blood or mucus from their hosts.

No one knows why there aren't more candirus in the world, but there may be some things about being a vertebrate that make a parasitic life hard. Vertebrates have high metabolisms compared with invertebrates, so they may not be able to get enough food within another animal. To be a parasite, an animal needs to produce a lot of young, because getting into the next host is so difficult and so essential. Vertebrates need to put a lot of energy into each offspring, so they may not be able to meet the challenge. But parasitism, as Richard Dawkins pointed out, doesn't have to take a conventional form like a tapeworm. Imagine an animal that could somehow trick another animal into raising its young. The tricker would be more likely to pass on its genes, while the trickee would have less time to tend to its own offspring and to its own genetic legacy. In fact, there are many species—both invertebrates and vertebrates—that practice just this sort of social parasitism.

Among the invertebrates, one of the most extreme cases can be found in the Swiss Alps. There you find nests of the ant *Tetramorium*. If you look for the queen, chances are good you'll find some pale, strangely shaped ants sitting on her back. They are not a special caste of *Tetramorium* ant but a different species altogether: *Teleutomyrmex schneideri*. *Teleutomyrmex* spends most of its life on a *Tetramorium* queen's back, hugging her with specially designed gripping legs. Instead of attacking these aliens, the *Tetramorium* workers let them eat the food they regurgitate for their queen. The *Teleutomyrmex* parasites mate inside their host's nest, and the new queens leave to find a new colony where they can hop on a new host.

The secret to parasitizing ants this way is creating illusions of smell. Ants depend mainly on smells to perceive the world, and they've evolved a complicated vocabulary of airborne chemicals to communicate with each other—to lay down food trails, to set off a colony-wide alarm, to recognize each other as nestmates. *Teleutomyrmex* can fool their hosts into caring for them rather than eating them because they can produce signals that make their hosts perceive them as queens themselves. The reason why *Teleutomyrmex* can cast these spells is probably that they evolved from their own host, turning their common language against their kin.

But many animals are social parasites of ants that aren't ants themselves. Some butterflies, for example, can trick ants into rearing their caterpillars. The butterflies lay their eggs on flowers, and when the caterpillars hatch, they drop to the ground, where ants come across them. Normally, ants look at a caterpillar as a gigantic lunch. But if they come across a social parasite, they act as if the caterpillar is a lost larva from own colony. Deceived by the caterpillar's odors, the ants drag it back to their nest, where they feed it and groom it the way they would any of their own larvae. Sometimes the ants even prefer the parasite to their own young. The caterpillar spends the winter growing in this luxury, after which it forms a cocoon. The ants go on caring for it as it metamorphoses into a winged butterfly. Only when it emerges from its cocoon does it finally occur to the ants that a huge intruder is in their midst and they try to attack it. But the butterfly bolts out of the nest and flies away.

All these social parasites essentially do what any conventional parasite does: they find the weaknesses in their hosts' defenses and turn them to their own advantage. There are verte-brates that do the same thing. The cuckoo, for instance, lays its eggs in the nests of other birds such as reed warblers. When a young cuckoo hatches, it proceeds to hurl its host's eggs and nestlings to the ground. The reed warbler feeds the cuckoo anyway, even as it grows so large that it dwarfs its stepparent. Once it is fully grown, the cuckoo flies off to find a mate, leav-ing the childless reed warbler behind.

Ants perceive their world mainly by smells, but birds de-pend much more on their eyes and ears. So cuckoos and other parasitic birds don't create fake smells but fake sights and sounds. The cuckoo egg mimics those of its host species, so the host is unlikely to get the urge to throw it out of the nest. After the cuckoo is born, it tricks the reed warbler into feeding by playing on the signals it uses to feed its young. To figure out how much food to catch, reed warblers look down in their nest, where their babies are holding open their mouths. If they see a lot of pink—the inside of bird mouths—they automatically hunt for more food. At the same time they rely on the sound of their crying babies as a second signal. If the babies are still hun-gry and are crying, the warbler will find more food.

A single cuckoo starts life much bigger than a warbler, and as it grows it gets even bigger. When the warbler looks down at its nest, it sees one big cuckoo mouth, which registers in its brain the same way a lot of little reed warbler mouths would. At the same time the young cuckoo mimics the calls of baby war-blers. But rather than mimic the sound of a single warbler, the cuckoo can sing like an entire nestful. So the cuckoo tricks its host not only into feeding it but into bringing it eight warblers' worth of worms. There may not be much room inside animals for a vertebrate parasite, but an animal's nest is another matter.

So is a mother's womb. When a fertilized egg tumbles down into the uterus and tries to implant itself, it encounters an army of macrophages and other immune cells. The new embryo doesn't have the same proteins on its cells as its mother, which ought to trigger the immune cells to destroy it. The fetus faces the same troubles as a tapeworm or a blood fluke, and it evades its mother's immune system in much the same way. The first cells that differentiate in a human embryo, known as trophoblasts, form a protective shield around the rest of its body. They fend off attacking immune cells and complement molecules, and they can send out signals that make the surrounding immune system sluggish. Strangely enough, there's some evidence that these suppressing signals are made in the trophoblasts by some of the viruses that are lodged permanently in our DNA—just as viruses in parasitic wasp genes let them control the immune systems of their hosts.

If you think of parasitism in terms of Dawkins's definition of genetic interests, then a fetus is a sort of half-parasite. It shares half its genes with its mother, and the rest belong to its father. Both mother and father have an interest, evolutionarily speaking, in seeing the fetus get born and live a healthy life. But some biologists have argued that parents also have strong conflicts on how the fetus grows. As it develops, it builds its placenta and a network of vessels to draw nourishment out of its mother. It knocks out its mother's control over her blood vessels near the uterus, so that she can't restrict the flow of blood to the fetus. It even releases chemicals to raise the concentration of sugar in her blood. But if the mother lets her child take too much, it might take a serious toll on her health. She might not be able to take care of her other children, and it might even threaten her ability to have any more. In other words, the fetus threatens her genetic legacy. Research suggests that mothers struggle against their fetus, releasing counteracting chemicals of their own.

While a fetus can take a heavy toll on its mother, how fast it grows will have no effect on its father's health. It's in his genetic interest for the fetus to grow as fast as possible. This conflict plays out within the fetus itself. Research on animals has shown that the genes a fetus inherits from its father and mother do different things, particularly in the trophoblasts. The maternal genes try to slow down the growth of the fetus, to control this parasite within her. Meanwhile, the paternal genes clamp down on these maternal genes and silence them, letting the fetus grow faster and draw more energy from its host.

Whenever two lives come into close contact and genetic conflict—even mother and child—parasitism will turn up.

· · ·

The feeling of being surrounded by a few million parasites is a hard one to put into words. If you put your face close to a jar filled with a graceful ribbon, a tapeworm pulled from a porcupine, you can't help admire its hundreds of segments, each with its own set of male and female sexual organs, all brimming with life and caught like a photograph in these preserving spirits. Then, just for a second, you start to worry that the whole creature will twitch a little, suddenly flail, and then break out of the glass.

The National Parasite Collection, run by the Agricultural Research Service of the U.S. Department of Agriculture, is one of the three biggest collections of parasites in the world. (Nobody is quite sure whether the American collection is bigger than the national collections of Russia. After you get up to a few million specimens, you tend to lose count.) It sits in a former guinea-pig barn on a farm the Department of Agriculture has been running in Maryland since 1936. In the distance, corporate headquarters push their cool blue-glass heads just over the trees. My guide through the collection was Eric Hoberg, a parasitologist in the shape of a bear. He studies the parasites of the far north, the nematodes that live only in the lungs of musk oxen, the flukes of a walrus. He led me down a flight of gray-striped stairs, past a couple of small labs, past a high stack of card catalogs a woman was slowly keying into a computer—a century of parasites. Then we went through a thick doorway to the collection.

At first I was a bit disappointed. I've followed paleontologists past museum displays and slipped through hidden doors into their collections, and we've wandered through corridors lined with high, deep cabinents full of whale skulls and dinosaur vertebrae that haven't been touched since they were dragged out of the ground. You could fit a little diner into the National Parasite Collection, or maybe a shoe repair shop. Hoberg introduced me to a retired science teacher named Donald Poling. Poling sat at a table, wearing hiking boots and a white lab jacket, rescuing slides of nematodes from preserving fluid that had crystallized over the past hundred years into the consistency of brown sugar. "Keeps me out of the bars," he said, scraping off a cover slip.

The rest of the room was taken up mainly by metal shelves on rollers that glided open with the turn of a three-pronged wheel. When Hoberg and I started walking among the shelves, browsing through the jars and vials, the disappointment disappeared. The collection surrounded me and became my world. We turned sealed jars around to read the labels that had been written in pencil. "Host: Yellowheaded Blackbird." Tapeworms from Alaskan reindeer. Liver flukes from elks. Frilly monogeneans that held on to the gills of fish from Korea.

At one point, when Hoberg was showing me a nematode—thick as a finger, long as a riding crop, the color of blood—which was still curled up inside a fox's kidney, I couldn't help myself. I said, "Gross." I had actually come to see Hoberg to learn something, not to continue with my horror marathon, but these things have a way of fighting their way out. Now it was Hoberg's turn for disappointment. "I get irritated by the yuck factor," he said. "What's being missed is how incredibly interesting these are. And it's tended to hurt parasitology as a discipline. Part of it is that people are put off by that," he nodded to the kidney. "Parasitologists are retiring and not being replaced by new ones."

We kept looking. We looked at a jar full of *Hymenolepis*, the

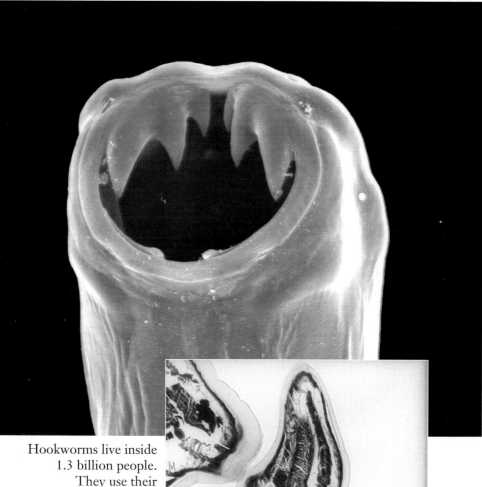

MALCOLM JONES, CENTRE FOR MICROSCOPY AND MICROANALYSIS, THE UNIVERSITY OF QUEENSLAND

Hookworms live inside 1.3 billion people. They use their powerful teeth to lacerate a patch of the intestinal wall (inset) and drink blood from the wound.

AFIP NEG. NO. 71-3163

Tapeworms, reaching up to sixty feet long, are the biggest parasites that live in humans.

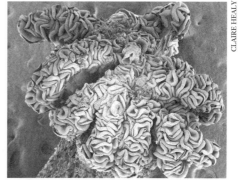

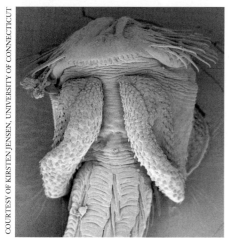

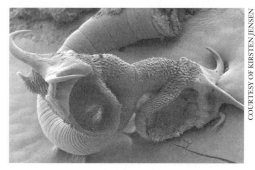

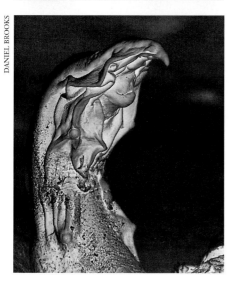

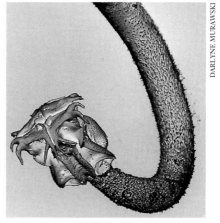

There are 5,000 known species of
tapeworm that live in various
animals, and probably many
thousands more still await
discovery. Each one has a head
specially adapted for lodging itself
in its host's body.

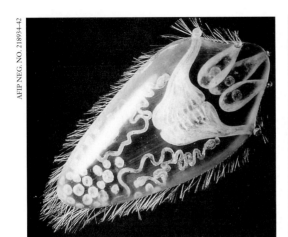

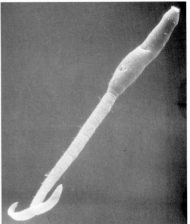

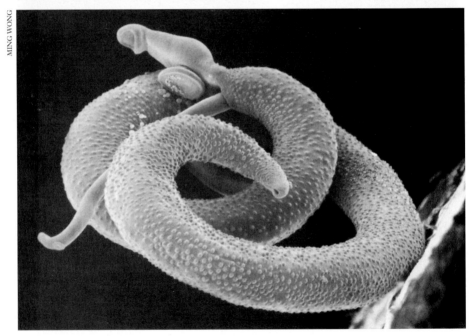

Schistosoma (also known as the blood fluke) infects more than 200 million people. *Top left:* Its eggs hatch in fresh water and the young parasite seeks out a snail. *Top right:* Inside the snail, the parasite passes through several generations before producing a missile-shaped stage called a cercaria. *Bottom:* The cercaria then penetrates human skin and becomes an adult that finally ends up in the veins of its human host.

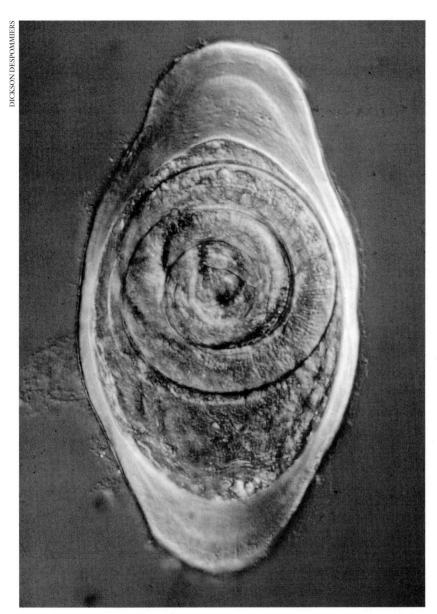

Trichinella, the cause of trichinosis, is an exceptional parasite: an animal that lives like a virus. Its larvae penetrate individual muscle cells and coil up inside, taking control of the muscle's DNA in order to make the cells a more comfortable home.

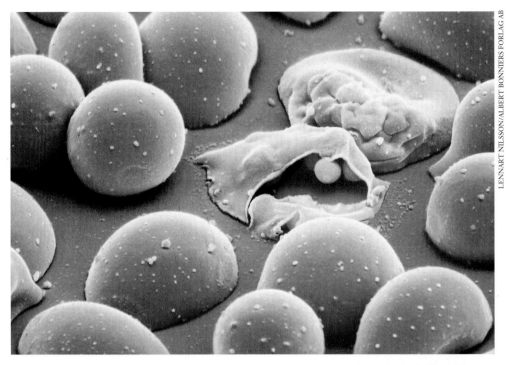

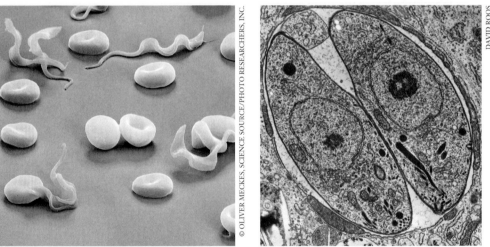

Top: The single-celled parasite *Plasmodium falciparum* causes malaria. Here a new generation of the parasite bursts out of a red blood cell. *Bottom left:* Another single-celled parasite, *Trypanosoma brucei,* is the cause of sleeping sickness. *Bottom right: Toxoplasma gondii* (shown here nestled inside a host cell) is one of the most successful parasites on Earth: in some regions of the world, 90 percent of people carry it in their bodies.

Parasites often choose very particular— and peculiar— places to live. This crustacean invades a fish's mouth, devours its tongue, and takes the tongue's place. It then acts like a tongue; the fish can use it to grip and swallow prey.

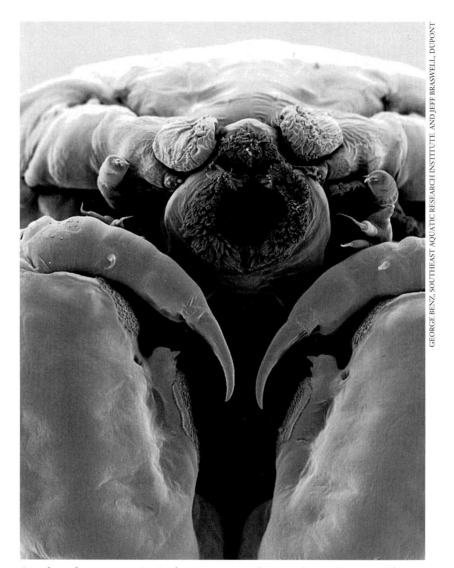

GEORGE BENZ, SOUTHEAST AQUATIC RESEARCH INSTITUTE AND JEFF BRASWELL, DUPONT

Another choosy parasite is the crustacean *Ommatokoita elongata*. It lives only in Greenland sharks, which roam underneath the Arctic ice. Moreover, *Ommatokoita* lives only in their eyes, anchoring itself in the eyes' jelly with its specially adapted legs.

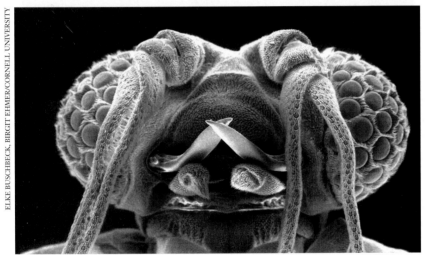

Insects are masters at parasitizing other insects. *Top:* Parasitic wasps lay their eggs inside caterpillars, and the larvae slowly devour their living hosts before crawling out and weaving cocoons. *Bottom:* The insect *Xenos peckii* makes non-parasitic paper wasps its host. When its eggs hatch, the female stays inside, devouring its hosts' sex organs, while the male burrows out and flies to another wasp to find a mate. As an adult, the male has only a few hours to live; as a result, it has evolved remarkable eyes to help find a mate. It has 100 miniature eyes, each of which is equipped with its own retina, able to form a full image of its own.

Once a parasite has used up its host, it needs to escape.
Top: A fungus emerges from an ant.
Bottom: A worm-like parasite called a nematomorph escapes its cricket host.

Thorny-headed worms, like many parasites, live in two or more hosts during their life. Many of them live initially in insects or crustaceans and then move into predators such as birds. To get into these predators, the parasites make their intermediate hosts stupid and foolhardy—and thus easily preyed upon.

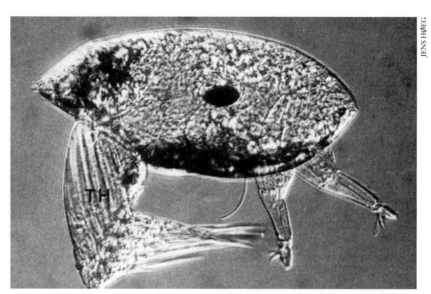

Top: The parasitic barnacle *Sacculina carcini* invades crabs and fills up their entire body with a network of roots. It forms a sac full of larvae where the crab's own egg pouch should be *(middle)*, and it forces the crab to care for its young. *Bottom:* Snails can also be horrifically victimized when they are infected with the fluke *Leucochloridium paradoxum.* The parasite's final hosts are birds. To get their attention, the parasite climbs into the snail's transparent tentacles. The striped flukes, which can be seen through the tentacles, look like caterpillars, and they catch the eye of hungry birds.

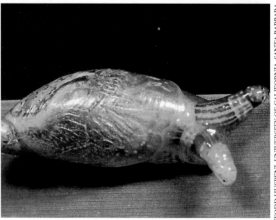

Cuckoos are a special sort of parasite—they don't live inside other animals, but they steal parental care. They lay their eggs in the nests of other bird species and trick the foster parents into rearing them. Here a reed warbler contemplates the giant cuckoo nestling that has taken its own offspring's place.

Only a few parasites of humans are on the verge of eradication. *Top:* For centuries people have extracted guinea worms from their legs by gently spooling them onto sticks. Public health campaigns have driven down guinea worms to less than 100,000 cases a year and are on the verge of eradicating the parasite altogether. *Bottom:* In 1998 a new campaign was launched to wipe out elephantiasis, caused by microscopic worms that block lymph nodes.

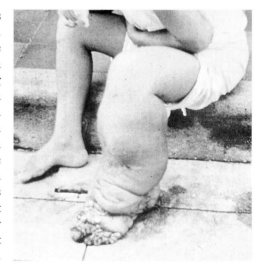

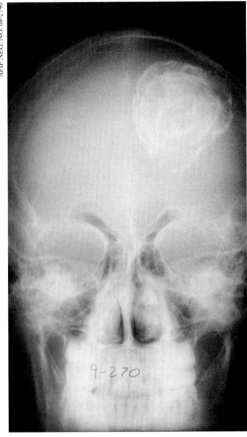

9-270

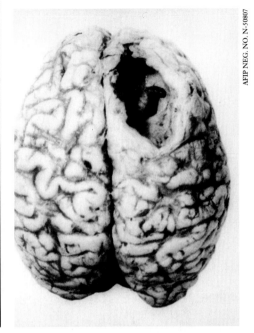

Parasites sometimes make mistakes, and the results can be lethal. *Top left:* Normally, tapeworms first mature in a cyst in intermediate hosts such as cows or pigs before they move on to humans. But if their eggs should end up in a human body, they will go ahead and form a cyst anyway, often in the brain. *Top right:* A botfly laid its eggs on a boy's head, and one of the larvae penetrated his brain. *Bottom:* Hosts have had to evolve ways to fend off the ever-present threat of parasites. Chimpanzees eat medicinal plants to fight off invaders.

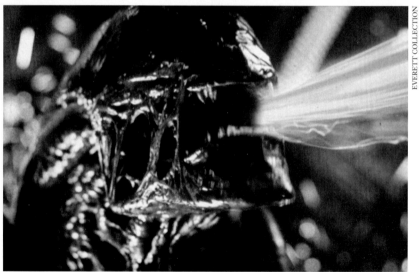

Hollywood has a healthy respect for the sophistication and cunning of parasites. *Top:* In the television show *The X-Files*, a fungus attacks people in the same way that some real fungi attack insects. *Bottom:* In the *Alien* movie series, a creature fashioned after parasitic wasps plants its young in the chests of human hosts.

tapeworm that uses beetles to get into rats, a great swirl of rice noodles. A piece of pig flesh with *Trichinella* running through it like a night of shooting stars. We passed closed trays of slides stacked upright like books on the shelves, hundreds of them, each with dozens of slices of parasites mounted on glass. We passed by the twelve thousand slides of specimens Hoberg collected in the Aleutian islands while he was working on his dissertation—twelve thousand slides he doubts he'll ever find time to write about before he retires. Hoberg brought the slides with him from the University of Washington when he got the job at the collection in 1989. A decade later, he was still coming across surprises. "Crab-eater seal?" he barked at a jar of tapeworms, picking it up and turning it in his hand. He lifted his glasses to his forehead to study the paper label floating in the fluid and said, "This may have been from Byrd's last expedition to the Antarctic." We came across a jar of botfly larvae. As horses walk through fields, adult botflies lay eggs on their hair, and when the horses lick themselves clean, they swallow the eggs. The eggs take the warmth of their mouth as a cue to hatch, and they chew their way into the horse's tongue. From there they drill down to the horse's stomach, where they anchor themselves and drink its blood. Once they mature, they let go their grip and are carried out of the horse's digestive tract. They hit the ground and transform into adult flies. In the jar before us, a swatch of horse stomach lay at the bottom, studded with botfly larvae, a cluster of stony little hives. I was fascinated, but Hoberg flinched. "That's *one* thing I can do without." I was glad to see that even a parasitologist has his limits.

Hoberg's favorite part of the collection was the slides. He grabbed a few cases and took them up with us to his office, which is dominated by a compound microscope. He focused slides for me to look at, showing sections of tapeworms from puffins, from bearded seals, from killer whales. It's hard to tell tapeworm species apart. Sometimes the only visual difference is the shape of the chamber that houses their sexual organs.

Sometimes only their genes will tell you that two tapeworms are separate species. Yet, by studying their relationships, Hoberg re-creates 400 million years of parasite history without a single fossil to guide him. He does so by finding strange patterns in parasites and their hosts. Why, Hoberg wonders, do these kinds of tapeworms—called tetrabothriids—live only in sea birds and marine mammals? Why do none of them live in humans or sharks? Why does another kind of tapeworm turn up in only two places in the world: in Australia and the thorn forests of Bolivia? The answers to these questions add up to a history of tapeworms, an epic that also carries secrets about the history of their vertebrate hosts, about drifting continents and pulsing glaciers.

A century ago, biologists thought this history was simple and drab. Once parasites surrendered to their inner life, they had reached an evolutionary dead end, since they could live nowhere else. What little evolution they experienced came only when their host dragged them in their wake. Their hosts might divide into new species when a population became isolated on an island or a mountain range, and the parasite, similarly cut off from the rest of its species, formed a new species of its own.

If that were true, you'd expect to see a certain pattern when you compared an evolutionary tree of closely related hosts to the parasites they carried: they would form mirror reflections of each other. Say you dissected four closely related bird species and found tapeworms inside. The lineage of birds that had branched off earliest on their own would have carried away the tapeworms that branch off first among the parasites. Each subsequent branch of host would have carried along its own branch of parasite as well.

It wasn't until the late 1970s that biologists such as Daniel Brooks of the University of Toronto started actually lining up host and parasite trees in this way. Before long they realized that these twinned histories were actually far more complicated

The Great Step Inward

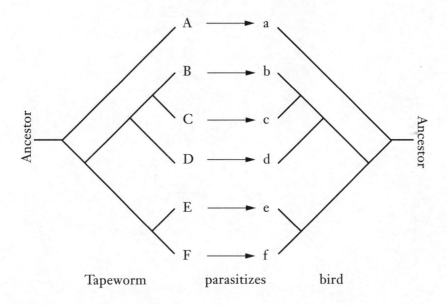

Tapeworm   parasitizes   bird

than they had thought. Sometimes the trees looked like perfect mirrors, like the tree above. But other times they looked like the tree on the next page.

Parasites did sometimes follow their hosts into new species, but they could also leap to entirely new hosts (as did tapeworms B, C, and E in this example). Sometimes they split into two new species on a single host without the host splitting as well. And sometimes they vanished from their hosts altogether. Parasites, in other words, have evolutionary stories as stormy and complex as their free-living cousins.

The most important clues to the early history of tapeworms come from the deepest roots in their tree. These primitive tapeworms all live in fish. Two main groups of fishes are alive today: the cartilaginous fishes, such as sharks and rays, and the bony fishes. They branched apart about 420 million years ago. About 400 million years ago, the bony fish lineage split into two branches of its own. One lineage led to ray-finned bony fish: salmon, trout, gar, and thousands of other species. The other

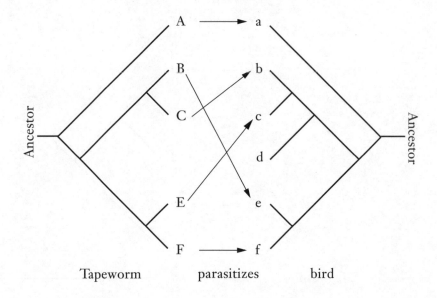

Tapeworm   parasitizes   bird

led to bony fish with fleshy lobe fins, such as lungfish and coela-canths. It was this lobe-finned branch that eventually produced vertebrates with legs, able to climb on shore—in other words, that became our ancestors.

Tapeworms probably first evolved in the earliest ray-finned fish. That history is reflected in the fact that the most primitive tapeworms alive today live in the most primitive ray fins, such as sturgeon and bowfin. It was in these hosts that tapeworms evolved from a leafy shape to their distinctively long, seg-mented bodies. From this origin, the tapeworms later colonized sharks and other cartilaginous fish. But apparently they didn't approach lobe fins. Neither lungfish nor coelacanths are known to carry the parasites.

Yet, tapeworms live inside their closest relatives—the ter-restrial vertebrates. In fact, they live in just about every sort of amphibian, bird, mammal, and reptile. Life on land didn't in-herit tapeworms from their aquatic ancestors. The parasites must have invaded them, coming out of the water in some ray-finned fish. Perhaps 50 million years after vertebrates had come ashore, some reptilian creature eating a fish picked up a tapeworm inside its meal, and a new lineage was born. Since then, tapeworms on land have evolved with their hosts as they diverged into new forms, and they've continued to hop from branch to branch, shuttling, for instance, from mammals to amphibians and from mammals to birds.

Vertebrates on land had split into reptiles and the forerunners of mammals by about 300 million years ago. By 200 million years ago, the reptile branch had produced dinosaurs, which rapidly became the dominant land animal. Did tapeworms live in dinosaurs? No one can say for sure, but it's hard to imagine they didn't, given that their closest relatives, birds and crocodiles, both carry them. And it's hard to imagine that they wouldn't have taken advantage of the space inside these giants, growing to lengths of one hundred feet or more. That's a thought that makes a parasitologist smile. The Santa Barbara parasitologist Armand Kuris has mused about what kind of ecology such a monster would have. The biggest dinosaurs were long-necked plant-eaters called sauropods, which could grow to weigh over one hundred tons. It's hard to fathom how any predator, even one as big as *Tyrannosaurus rex*, could have brought them down. Perhaps it only scavenged the big dinosaurs, or perhaps it got some help. Perhaps, Kuris has suggested, the tapeworms turned the sauropods and *Tyrannosurus rex* into foreshadowings of moose and wolf. The sauropods picked up tapeworm eggs on the plants they ate, and the parasites developed into giant cysts inside them. As they tore up their hosts' lungs or brains, they might have slowed down the sauropods enough to let *Tyrannosaurus rex* catch them, and let the tapeworm get into its final host. A dinosaur tapeworm might even have left its mark on the fossil record. The cysts of some tapeworms today get so big, and grow with such force, that they can split open a human skull. If dinosaurs carried cysts so big you'd need a forklift to carry them, paleontologists might be able to recognize their traces.

Over the 400 million years that tapeworms have been alive, Earth has been blasted by four major mass extinctions. The most recent one took place 65 million years ago and was most likely triggered by a ten-mile-wide asteroid that crashed into the Gulf of Mexico. It was powerful enough to kill the dinosaurs as well as 50 percent of all species on Earth. But tapeworms survived. It's even possible in some parts of the world to find tapeworms still living the way they did when dinosaurs walked the Earth. The thorn forests of Bolivia are home to marsupials such as mouse opossums. They are hosts to a rare group of tapeworms called linstowiids, which need an arthropod as an intermediate host. The only other place on Earth where linstowiid tapeworms live is Australia, where they also live in similar marsupials. Today these parasites are split by thousands of miles of Pacific water, but 70 million years ago, Australia, South America, and Antartica were all joined in a single continental mass. The ancestor of the Australian and Bolivian tapeworms originated in a marsupial on that vanished continent, and host and parasite gradually split apart as the land mass was split by continental drift. But over the 70 million years that have since passed, the ecosystem that supported the tapeworm's cycle through the mammals has remained intact.

Other tapeworms may have survived the asteroid by abandoning their old hosts. The tetrabothriid tapeworms live only in marine birds like puffins and grebes, and marine mammals like whales and seals. On the face of it, this sort of combination of hosts doesn't make sense. These animals are too distantly related to share the tapeworms as an heirloom from some common ancestor. Birds evolved from reptiles—probably ground-running dinosaurs over 150 million years ago. Marine mammals invaded the oceans much later. Whales arose from coyote-like mammals about 50 million years ago, and seals from bear-like mammals about 25 million years ago. You have to reach back over 300 million years to find a common ancestor for birds and mammals, and that same ancestor gave rise to many other lineages of vertebrates, ranging from crocodiles to tortoises to cobras to wallabies to humans—none which is a host for tetrabothriids.

The birds and the whales had to get their tapeworms from somewhere. They probably didn't get them from fish, because the closest relatives of tetrabothriids live in reptiles on land, which aren't closely related to the birds and the whales. So tetrabothriids must descend from a tapeworm that lived in some group of ancient reptilian hosts. It just so happens that before whales and sea birds existed, there were reptiles in the oceans that played the same ecological roles. If you had sailed across an ocean 200 million years ago, you wouldn't have seen birds flying overhead but pterosaurs: narrow-headed reptiles that soared on wings of hairy skin, plucking fish to bring back to their rookeries on shore. And breaching the water around you would not have been whales but monstrous reptiles of many pedigrees, such as long-necked plesiosaurs and sword-fish-shaped ichthyosaurs.

Between 200 and 65 million years ago, these reptiles dominated the marine food chain. Pterosaurs began sharing the sky with birds, and Hoberg thinks that as a sort of welcoming present, they gave them their tapeworms as the birds ate the fish that served as the parasite's intermediate host. The extinction 65 million years ago that claimed the big dinosaurs also wiped out the marine reptiles and the pterosaurs. No one knows why birds survived the impact, but it seems that they carried on the cycle of the tetrabothriid. Whales and seals later took up the roles left vacant by the marine reptiles, and the tapeworms colonized them as well. As long as an ecosystem remains intact—even if the animals that constitute it change—parasites will survive.

In the past 65 million years, tapeworms have continued to thrive, and their travels continue to mark the history of their hosts. The tapeworms that live in stingrays in the Amazon, for example, show how the river once flowed backward. If stingrays had colonized the Amazon from the Atlantic, where it flows today, their tapeworms would be most closely related to tapeworms in living Atlantic rays. But the tapeworms are actually more closely related to those in the Pacific. And making matters more puzzling, there are still other tapeworms in the Atlantic and Pacific stingrays that are more closely related to one another than either is to the Amazon tapeworms.

The scenario that reconciles these facts best has stingrays coming upriver 10 million years ago. At that time, the Andes hadn't yet formed, and the Amazon flowed out of Brazil to the northwest coast of South America. Another big difference in the geography of that time was that the isthmus of Panama hadn't yet formed, so that the Atlantic and Pacific were joined by a broad channel. Groups of stingrays from the Pacific swam into the Amazon when it flowed in the opposite direction. As the Amazon stingrays adapted to fresh water and became isolated from their ocean-going cousins, the marine stingrays still mingled between the two oceans. By the time Panama had risen out of the ocean, they had shared some new species of tapeworms that the freshwater rays couldn't pick up.

In the last few million years, tapeworms have discovered yet another host, one that walks on two legs. Hoberg has been studying tapeworms that live in humans. Over the years parasitologists have come up with many ideas for how tapeworms came to live inside us. One has it that ten thousand years ago, when humans domesticated livestock, they picked up the tapeworms that cycled between wild relatives of cattle and their predators. But looking at evolutionary trees, Hoberg doesn't think that's the case. He and his colleagues have compared the genes of human tapeworms with their closest relatives and have found they branched off on their own a million years ago, not a few thousand. At that point, our ancestors were hominids who were a long way from farming. The closest thing to a cow or a pig they would have eaten then would have been the scavenged carcasses of wild game that had been killed by lions. Which would explain something else Hoberg discovered: the closest relatives to human tapeworms make lions and hyenas their final host. Hoberg pictures hominids following after lions, scavenging their kills and picking up their tapeworms.

There is more than one way to look back at the dawn of humanity. You can go to Ethiopia and sift the dust for stone tools and scoured bones, but you can also go to the National Parasite Collection, find the right jar, and stare at a fellow traveler.

• • •

As tapeworms moved into new hosts they had to evolve new ways to live inside them. They had to adapt to new geographies of intestines; the tapeworms that began living inside rats stumbled across new ways to get flour beetles into their final host's jaws. Reconstructing the rise of these adaptations is treacherous work because sensible-sounding stories about evolution are easy to make up. You see long tails on a swallow and decree that they must have evolved to let the bird maneuver more precisely, but someone else looks at them and decrees that they have evolved that way because female swallows find them attractive on male ones. Or maybe no adapation is involved at all—maybe most of the swallows that happened to establish this species just happened to have long tails, and it's been that way ever since.

Consider the journeys of the nematode *Strongylus*. In one species, for instance, *Strongylus vulgaris*, the larva crawls to the top of blades of grass and lies in wait for a horse to graze by. Once swallowed, the worm takes a long, seemingly pointless journey. It travels down the horse's throat to its stomach and then passes on into the gut. From there it chews out into the horse's abdominal cavity and wanders the arteries of the horse for weeks until it has matured. Thereupon it returns to the intestines, burrows its way back in, and spends the rest of its life there.

Why should a parasite leave the intestines only to return for the rest of its life? Suzanne Sukhdeo has sorted through the close relatives of *Strongylus* and she has come to a working hypothesis for how this pilgrimage came to be. The ancestor of these nematodes lived in the soil more than 400 million years ago, spending its days burrowing and feeding on bacteria, amoebae, and other microscopic game (as many thousands of species of nematodes still do today). About 350 million years ago, it began to encounter something new—soft-skinned amphibians slithering around in the muck. The nematodes used their burrowing abilities to plow into these hosts and make their way to the gut, where they lived happily on the food that the amphibians ate.

Over the course of tens of millions of years, new kinds of vertebrates evolved on land: upright mammals and reptiles. These animals no longer offered the easy target of a slimy belly hugging the ground—they stood high on tall legs. Some parasitic nematodes adapted to these new hosts by evolving a new entry: by getting eaten rather than burrowing in through the skin. But burrowing, Sukhdeo argues, was too deep in their nature to disappear. Once swallowed, they would take up the flesh-drilling pilgrimage their ancestors had made for millions of years, looping back through their host's body in order to enter the intestines again.

Sukhdeo suggests that the strange trip of *Strongylus* is just an evolutionary relic. Some day they may lose this heritage, but for now they still retain a vestige of their first go at parasitism, when bellies and mud stayed in close touch. On the other hand, some researchers think the parasites continue taking this journey because it benefits them. Parasitologists have compared species of nematodes such as *Strongylus* that wander through tissue with species that stay put in the intestines, and they've found a pretty consistent difference: the wanderers actually grow faster and end up bigger and more fertile. A trip through muscle means a respite from the gastric acid of the intestines, the slosh of digested food, the low oxygen levels, and the vicious blasts of the intestine's powerful immune system. The trip may be a relic, but it's a useful one.

The puzzle of parasite evolution gets even more confusing when you consider the things that happen to hosts when they are invaded by parasites. Filarial worms, which cause elephantiasis, enter the lymphatic system and start producing thousands of baby worms. Sometimes a person's immune system reacts violently to the worms, scarring the lymph channels and blocking them up. The lymphatic fluid builds up in the lymph channels, producing elephantiasis—monstrously swollen legs, breasts, or scrotums. There'd be no sense in calling a swollen leg an adaptation of the parasite, since it does no good for the worm. It's simply the immune system misfiring. It is nothing more than what Richard Dawkins has called a "boring by-product."

The best way to tell whether a given change to a host is a boring by-product or a true adaptation is to study its evolution. One elegant test of this has been done with insects that make galls on plants. You may sometimes notice cherry-shaped balls hanging from the leaves of oak trees, or a flower's stem bulging as if it had somehow swallowed a marble. These are galls: bits of plant tissue that have formed into shelters for insect parasites. Hundreds of different insect species live in galls, which can form on flowers, twigs, stems, or leaves. Some species of wasps, for example, lay their eggs on oak leaves, and the cells of the leaf respond to the egg by growing up and around it. The larva is born and becomes buried even deeper in the leaf. The cells multiply into a huge spherical shape, with an inner layer of hairy tissue. Food—starches and sugars, fats and proteins—is pumped into the gall from elsewhere in the plant and fills up the oversized cells in the inner hairs. The wasp larva bursts them open and feeds on the fluid cocktail. As it destroys the inner cells the outer ones divide and become ready to be eaten.

The galls are formed by the plants themselves, not the insects. Are they, as some researchers have suggested, just scars that happen to give the parasites some shelter? Warren Abrahamson of Bucknell University and Arthur Weis of the University of California at Irvine have performed some of the closest studies of galls, focusing on the goldenrod gallflies. The flies lay their eggs in a bud of a goldenrod plant in late spring. A spherical gall forms, growing to half an inch to an inch in diameter, and the fly larva grows inside. Parasitic wasps attack the fly larva, as do beetles. Woodpeckers and black-capped chickadees chip the galls open during the winter to eat them like some kind of delicious hard-shelled nut.

The galls in which these flies live vary in size and shape. Say for the moment that the galls are merely the boring by-product of a fly living within a goldenrod plant. Then you'd expect that any change in their variation from one generation to the next should be linked to changes in the genes plants use to defend themselves against invaders. Abrahamson and Weis have run experiments in which they raised gallflies on goldenrod plants that were all clones. Since their genes were identical, the plant's defense against the flies should have been identical. Yet, Abrahamson and Weis found that the plants produced very different sorts of galls. That suggests that the flies' genes are responsible for shaping the galls by taking control of the plant's own genes. There's probably some fierce natural selection going on in the flies for these genes, given that 60 to 100 percent of the galls are attacked by parasites. Supporting this, when the biologists observed the gallflies from generation to generation, a given lineage of flies all produced similar galls. The gall is made by the plant and yet is the work of the parasite, shaped by its evolution, not that of its host.

It's actually surprising just how many things parasites do to their hosts that are *not* boring by-products but adaptations produced by evolution. Even harm itself is often an adaptation. Closely related parasites can be gentle or brutal to their hosts, or any shade in between. *Leishmania* can cause a few sores or eat away your face, depending on the species. Until recently, scientists didn't think about how parasites could have such different effects on their hosts. The doctors were too busy looking for cures, and the evolutionary biologists were more interested in hosts than in parasites. They waved off the differences with a notion that when parasites first hop to a new host species they do a lot of damage. Once they've had a chance to fine-tune themselves, the story went, the parasites gradually mellow.

That's certainly the case when many parasites accidentally find themselves in new hosts. A disease called sparganosis, for example, is caused by a species of tapeworm that uses copepods as its intermediate host and matures inside a frog. If a human should accidentally swallow the copepod in a glass of water, the tapeworm will escape out of the intestines and wander in confusion around the body, with none of the cues and landmarks it uses in a frog. As it zigzags randomly under the skin the tapeworm grows a few inches long, destroying tissue in its wake and inflaming its host into agony. If enough frog tapeworms found themselves inside humans, they might evolve into a new species better adapted to a new host. If they did, the conventional wisdom went, they would be amply rewarded by natural selection for any mutation that caused less harm to their new host. After all, if their host died off, the parasites would die with it. The wisdom of maturity brings gentleness.

It took until the 1990s for biologists to run the first experiments that could actually test this notion. A German evolutionary biologist named Dieter Ebert performed one of them, using water fleas. Water fleas sometimes suffer from a parasitic protozoan called *Leistophora intestinalis*, which lives in their gut and gives them diarrhea; the diarrhea carries the parasite's spores with it, spreading them to other water fleas in the same pond. Ebert gathered fleas from England, Germany, and Russia and raised parasite-free colonies of each population. He then infected the colonies with *Leistophora* but used only the ones that had lived in the English ponds.

According to the conventional ideas about parasites, the English water fleas should have fared best. After all, the English

Leistophora had spent untold generations inside the English water fleas and theoretically had come to a mellow coexistence. But Ebert found in fact that the opposite happened. The English fleas became burdened with many more parasites than the German and Russian fleas: they grew more slowly, they laid fewer eggs, and they died in greater numbers. Even though the English parasites had had more time to adapt themselves to English fleas, they had remained vicious.

Ebert's findings did not come as a surprise to some biologists. They had built mathematical models of the relationship between hosts and parasites, and they had discovered theoretical reasons why familarity could breed contempt. Natural selection favors genes that can get themselves replicated more often than others. Obviously, a gene that makes a parasite instantly fatal to its host won't go very far in this world. Yet, a parasite that is too well mannered won't have any more success. Because it takes almost nothing from its host, it won't have enough energy to reproduce itself and will come to the same evolutionary dead end. The harshness with which a parasite treats its host— what biologists call virulence—contains a trade-off. On one hand, the parasite wants to make use of as much of its host as possible, but on the other hand, it wants its host to stay alive. The balancing point between these conflicts is the optimal virulence for a parasite. And quite often, that optimal virulence is quite vicious.

The way virulence works is nicely illustrated by mites that live on the ears of moths. Moths have to be on constant guard against bats, which seek them out with echolocating shrieks. When moths hear the bats sending out their ultrasonic signals, they immediately start dodging and weaving through the air to avoid an attack. If the mites colonize the full extent of a moth's ear—on both its outside and its inside—they will have enough room to produce a lot of offspring. But as they root around, damaging the delicate hairs that the moth uses to hear, they leave the moth deaf in that ear. With one ear out of commission, the moth will have a harder time escaping bats. If both ears shut down, the moth is doomed.

Nature has settled on two solutions to this dilemma. Some species of mites take up residence in the entire ear, both on the outside and on the inside. But they live in only one of the moth's ears, leaving their host with enough hearing to keep it from being devoured. Other species of mites live on the outside of both ears. But because they forgo all the inner-ear real estate, they reproduce less than the deafening mites and are transmitted more slowly from moth to moth.

To test theories of virulence, biologists can make predictions about how real-world parasites behave. In the forests of Central America, several species of parasitic nematodes live inside wasps. These wasps are exceptional creatures: the female lays her eggs inside the flower of a fig tree and dies. The flower transforms into a plump fruit, and the eggs of the wasps hatch, the wasp larvae feeding on the fig. They mature into adult males and females and mate inside the fruit. The females then leave the fig to find a new one to lay their eggs in. As they leave they gather pollen on their bodies, and when they find a new fig flower, they fertilize it, triggering the production of a new seed.

It's a pleasant symbiosis for both plant and animal: the fig depends on the wasp to let it mate, and the wasp depends on the fig for a place to raise its young. But into this happy scene intrudes the nematode. Some figs are riddled with these parasites, and when an egg-bearing female wasp prepares to leave, a nematode crawls onto her to hitch a ride. By the time the wasp has arrived at a new fig, the nematode has penetrated her body and is devouring her guts. The wasp enters the fig and lays her eggs, but the parasite has laid its own eggs inside her body as well. By the time the wasp has finished laying her eggs, the parasite kills her, and out of her body emerge a half dozen or so new nematodes.

The wasps and nematodes have been living together as host and parasite for over 40 million years—a long, venerable association. From species to species the wasps have different egg-laying habits: some will lay eggs only in a fig untouched by other wasps so that their young will have the fig to themselves. Other species don't mind laying eggs alongside those of other wasps. Virulence theory makes a prediction about the nematodes that live in fig wasps. Nematodes that infect a wasp that lays its eggs alone must handle their host delicately. If they ravage the wasp too quickly, she may be able to lay only a few eggs, or none at all. The nematode's own offspring would then have fewer potential hosts in their fig, and they'd have worse chances of surviving.

The same doesn't hold for parasites of more neighborly wasps. When a nematode's offspring hatch in a fig, they're likely to find other wasps there that they can parasitize. What a nematode does to its own host doesn't pose any risk to its offspring, so you'd expect these parasites to be far nastier. The biologist Edward Herre studied fig wasps and their parasites in Panama for over a decade, and when he looked over his records for eleven species, he found that they did indeed fall into the predicted pattern—a powerful vindication for the theory of virulence.

To study the laws of virulence, parasitologists can work with just about any parasites, whether they are mites, nematodes, fungi, viruses, or even rogue DNA. The host can be a human, a bat, a wasp, an oak tree. The same equations still apply. When scientists look at parasites from this evolutionary point of view, suddenly the walls that traditionally divide them tumble away. Yes, they all occupy different branches of the tree of life; yes, they are all descended from radically different free-living ancestors. But those gulfs make their similarities all the more remarkable. Darwin himself noticed that different lineages can independently evolve toward the same form. A bluefin tuna and a bottlenose dolphin are separated by over 400 million years of divergent evolution. Yet, the dolphin, whose ancestors looked like coyotes only 50 million years ago, has evolved a teardrop-shaped body, a rigid trunk, and a narrow-necked tail shaped like a crescent moon—all of which are possessed by the tuna. Biologists call this coming-together convergence, and parasites are the most spectacularly convergent organisms of all. Free-living nematodes have moved from the soil into the roots of trees, where they have evolved the ability to switch on and off individual genes and turn individual plant cells into comfortable shelters. Another lineage of nematodes produced *Trichinella*—a parasite that does the same thing to the cells in muscles of mammals. The lancet fluke has evolved chemicals that can force an ant to climb to the top of a blade of grass and clamp itself there. The same feat is accomplished by fungi. To find the last common ancestor of lancet flukes and fungi, you'd have to explore the oceans for some single-celled creature that lived a billion years ago or more. Yet, after all that time, they both managed to come across the same tactic to control their hosts.

The laws of virulence are also built on convergence, and they promise to change the way we fight diseases. A virus such as HIV needs to go from host to host in order to propagate, just as a nematode does. If it becomes easier for a strain of HIV to travel, it can reproduce more quickly in a given host (and cause him or her more harm). That's how the AIDS epidemic has played out: in populations where people have many sexual partners, the virus destroys its host's immune system faster. Cholera is caused by a bacterium called *Vibrio cholerae*, which travels through water and escapes its host by causing diarrhea. In places where water is purified and *Vibrio*'s odds of infecting a new host are low, the disease is milder. In places without sanitation, the bacteria can afford to be more vicious.

The history of parasites, stretching over billions of years, is just beginning to emerge, but already it has made clear that degeneration isn't its guiding force. Parasites may indeed have lost some traits over the course of their evolution, but then again, in our own history we have lost tails, fur, hard-shelled eggs. Lankester was appalled at how *Sacculina* lost its segments and appendages as it matured. He could just as easily have been disgusted by the way he himself had developed the vestiges of gills in his mother's womb and then lost them as he grew lungs. As parasites colonized Earth's third great habitat, they did lose some of their old anatomy, but they evolved all sorts of new adaptations that scientists are still trying to understand.

At the end of the day at the National Parasite Collection, after Eric Hoberg and I had spent an afternoon in his office talking and searching slides, I asked if I could go back down to the collection. "Sure. Just let me unlock it for you," he said. We walked back downstairs, and he opened the door. It was empty now; Donald Poling had finished his slide-scraping for the day and had gone home. As I walked in, Hoberg stood by the door and told me to ask if I needed anything, and then he shut me in. The heavy door closed with more finality than I would have liked. Now I was trapped with the parasites. But after I got used to being shut in with them, the place became meditative. This was the closest thing to a proper museum I could think of for parasites, even though a grand diaspora of parasites was missing—the parasitic wasps and gall makers scattered in entomological collections, the protozoa hidden in schools of tropical medicine, *Sacculina* in the hands of some Danish expert on barnacles. Someday, I thought, you'll all be reunited, and maybe in something classier than an old guinea pig barn.

6

Evolution from Within

The wise learn many things from their enemies.
                    —Aristophanes, *The Birds*

The Origin of Species is a mournful book. God did not put species here on Earth balanced in perfect harmony, Darwin was saying. They are born out of a vast, ongoing death. "We behold the face of nature bright with gladness, we often see superabundance of food," he wrote. "We do not see, or we forget, that the birds which are idly singing round us mostly live on insects or seeds, and are thus constantly destroying life; or we forget how largely these songsters, or their eggs, or their nestlings, are destroyed by birds and beasts of prey." Most plants and animals never get a chance to reproduce, he argued, because they are killed by some predator or grazer, are outcompeted by members of their own species for sunlight or water, or just starve to death. The few that survive all these menaces and reproduce pass on their secret to success to the next generation. And out of all this death comes natural selection, which can transform it into the songs of birds, the leap of a flying fish—into a world that looks, at least on its surface, bright with gladness.

Yet, Darwin said little about one particularly powerful evolutionary menace, one that brought him a lot of personal sadness. His ten children struggled against diseases such as influenza, typhoid, and scarlet fever, and by the time *The Origin of Species* came out in 1859, three of them had died. Darwin himself suffered for much of his adult life with fatigue, dizzy spells, vomiting, and heart trouble. He once described his health this way: "Good, when young, bad for the past 33 years." Although no one is sure what made him suffer, some have suggested that he had Chagas disease. Chagas disease is caused by *Trypanosoma cruzi*, a species of trypanosomes related to *Trypanosoma brucei*, the cause of sleeping sickness. *T. cruzi* slowly wrecks parts of the nervous system, and the ways to die of Chagas are horrible in their variety: your misfiring heart may stop beating, for example, or your intestines may stop getting the proper commands for peristalsis and let food pile up in the colon until you die of blood poisoning. *T. cruzi* is spread by the benchuca, a biting insect of South America, and Darwin was bitten by one as he was traveling the world on the *H.M.S. Beagle*; many of his symptoms arose only when he returned to England. The Darwins didn't have to worry about getting eaten by wolves or starving to death, but infectious diseases—in other words, parasites—could still ravage them.

The toll that parasites take on the rest of life is far heavier— a toll that in terms of evolution is on par with predators and starvation. Viruses and bacteria tend to do their work quickly, multiplying madly and causing diseases that either kill or are defeated by the immune system. Eukaryotic parasites can be swiftly fatal as well—witness the brutality of sleeping sickness and malaria—but they can also do other kinds of damage. Ticks and lice may only live on the skin, but they can leave their host gaunt and emaciated. Intestinal worms can let their hosts live for years, but they stunt their growth and cut down their litters. The flukes that Kevin Lafferty studied in the Carpinteria salt marsh don't destroy their killifish hosts themselves, but they turn them into dancing bird food. A crab infected with *Sacculina* may live a long life, but because it has been castrated by its parasite, it cannot pass on its genes. Evolutionarily speaking, it's a walking corpse.

By keeping their hosts from passing on their genes, parasites create an intense natural selection. Perhaps parasites caused Darwin too much misery for him to recognize that they can be a creative evolutionary force in their hosts. A lot of the evolution that results takes place where you'd expect it: in the immune system, which defends animals from invaders. But it also brings out things that seem at first to have nothing to do with diseases. There's growing evidence that parasites are responsible for the fact that we, and many other animals, have sex. The tail of a peacock, and other devices that males use to attract females, may be brought to us thanks to parasites. Parasites may have shaped societies of animals ranging from ants to monkeys.

Parasites have probably been driving the evolution of their hosts since the dawn of life itself. Four billion years ago, when genes formed loose confederations, parasitic genes could take advantage of them and get themselves replicated faster than the rest. In response, these early organisms probably evolved ways to police their genes. This sort of monitoring still goes on today in our own cells, which carry genes that do nothing but search for genetic parasites and try to suppress them.

When multicellular organisms evolved, they became a particularly choice target for parasites, since each one offered a big, stable habitat rich with food. And multicellular organisms had to fight a new sort of parasitism as well, in which some of their own cells tried to replicate at the expense of the rest of the organism (a problem we still face with cancer). All these pressures led to the evolution of the first immune systems. But for every step that a host takes against parasites, parasites are at liberty to evolve a step in response. Say an immune system evolves a tag it can put on parasites to make them more recognizable and easier to kill. The parasite can then evolve the tools it needs to rip that tag off. Immune systems became increasingly sophisticated in response; about 500 million years ago, for example, vertebrates evolved the ability to recognize specific kinds of parasites with T and B cells, and make antibodies to them.

This evolutionary back-and-forth didn't just happen back in the depths of time. It happens today, and biologists can watch it in action if they run the right sort of experiment. A. R. Kraaijeveld of the Imperial College in England performed one such experiment with fruit flies and the wasps that parasitize them. For his experiment, he chose a wasp and two of its host species: the fruit flies *Drosophila subobscura* and *Drosophila melanogaster.* He raised the wasps on *D. subobscura* flies, and then put a few dozen of the parasites in a chamber with *D. melanogaster.* The wasps parasitized these new hosts, and they killed nineteen out of every twenty *D. melanogaster.* But one out of twenty *D. melanogaster* managed to marshal its immune system and kill the wasp larvae. Kraaijeveld took these resistant fruit flies and used them to breed the next generation of *D. melanogaster.*

Meanwhile, Kraaijeveld continued to raise his wasps on the other flies, *D. subobscura.* When the next generation of *D. melanogaster* had matured, he took some of the wasps and transferred them to their chamber. The wasps would then attack the new generation of *D. melanogaster,* and once again, Kraaijeveld would raise the survivors to produce a new generation.

By raising the wasps and the flies in this way, Kraaijeveld was blindfolding one of the boxers in his host-parasite match. With each generation, the *D. melanogaster* flies were able to adapt more and more to the wasps. But the wasps, which Kraaijeveld raised on another species of fly, didn't have any chance to match the evolution of their *D. melanogaster* host. The mismatch let *D. melanogaster* steadily improve their fight against their parasites. In only five generations, the proportion of flies that could kill the wasp larvae rose from one in twenty to twelve in twenty.

Hosts and parasites may evolve together in a continual escalation (what biologists call an arms race), but in many cases their evolution can look more like a merry-go-round. Parasites evolve over time to do a better and better job of recognizing their hosts, finding weakness in their defenses, and thriving inside them. But a host species is never genetically uniform—it instead comes in strains, each with its own set of genes. Parasites have variations of their own, and some of them may help parasites against particular strains of hosts. Over time strains of parasites emerge, each adapted against strains of hosts.

Biologists have built mathematical models of these intimate relationships. If one strain of host is more common than the rest (call it Host A), any parasites that are adapted to it will have a rosy future. After all, they can hop between a wealth of hosts, replicating along the way. The problem is that, as parasites, they will kill or disable a lot of their hosts. From generation to generation, Host A will fade as its parasites undermine their success.

The attention that parasites pay to the most common host gives rarer host strains an advantage. Since the most common parasites aren't adapted to attack them, they get the opportunity to multiply. As Host A declines, another host, say Host B, rises. But then parasites that can adapt to Host B get rewarded by natural selection and multiply as well. They eventually drive down Host B's numbers, letting Host C ascend, then D, and E, and so on, maybe even back to Host A again. Every now and then a mutation creates a rare new strain of host. It simply becomes Host F and falls into the rotation.

This endless rise and fall would probably have appalled the biologists of Lankester's day. They saw the history of life as a march of progress, always threatened by degeneration. In this new kind of evolution there is no progress forward or backward. Parasites force their host to go through a huge amount of change without going anywhere in particular. One variant rises, then it falls, and another variant rises to take its place, only to fall in turn. This sort of evolution isn't the stuff of epic poetry but of surreal children's stories. Biologists came to call it the Red Queen hypothesis, referring to the character in Lewis Carroll's *Through the Looking Glass* who took Alice on a long run that actually went nowhere. "Now, *here*, you see, it takes all the running *you* can do, to keep in the same place," the Red Queen declared.

Yet, there's a paradox to the Red Queen hypothesis. While it's all about running to stay in place, it may have allowed evolution to take one crucial step forward: it may have brought about the invention of sex.

• • •

In the early 1980s, Curtis Lively found himself in New Zealand wondering about sex. He had just finished earning a Ph.D. in evolutionary biology by studying the barnacles of the Gulf of California. One of the questions he had to answer on his qualifying exams was, Why did evolutionary theory have such a hard time accounting for sex? He had no idea.

It's not a question that most people are accustomed to asking. "If you go into a class of sophomores and ask, 'Why are there males?,' they look at you as if you're crazy," Lively says. "They'll say you need males to reproduce and that each generation produces more males. Well, that may be true for mammals, but for many species that's not true. It's just staggering to them to think that anything could do that, could reproduce without males and sex. Sex and reproduction are just fused in most people's brains."

Bacteria simply divide themselves in two when the time seems right, as can many single-celled eukaryotes. Many plants and animals have the ability to reproduce themselves on their own quite comfortably. Even among the species that do reproduce sexually, many can switch over to cloning. If you walk through a stand of hundreds of quaking aspen trees on a Colorado mountainside, you may be walking through a forest of clones, produced not by seeds but by the roots of a single tree that come back up out of the ground to form new saplings. Hermaphrodites, such as sea slugs and earthworms, are equipped with male and female sex organs and can fertilize themselves or mate with another. Some species of lizards are all mothers: in a process called parthenogenesis, they somehow trigger their unfertilized eggs to start developing. Compared with these other ways to reproduce, sex is slow and costly. A hundred parthenogenetic female lizards can produce far more offspring than fifty males and fifty females. In only fifty generations, a single cloning lizard could swamp the descendants of a million sexual ones.

When Lively was learning about the mystery of sex, there were only a handful of good hypotheses to explain why it existed at all. Two of the favorites were nicknamed the Lottery and the Tangled Bank. According to the Lottery hypothesis, sex helped life survive in unstable environments. A line of clones might do well enough in a forest, but what if that forest changed over a few centuries to a prairie? Sex brought the variations that could allow organisms to survive change.

According to the Tangled Bank hypothesis, on the other hand, sex gets offspring ready for a complicated world. In any environment—a tidal flat, a forest canopy, a deep-sea hydrothermal vent—the space is divided into different niches where different skills are needed for survival. A clone specialized for one niche can give birth only to offspring that can also handle the same niche. But sex shuffles the genetic deck and deals the offspring different hands. "It's basically spreading out progeny so that they're using different resources," says Lively. The progeny wouldn't have to fight with each other over food as much, and thus a mother would be more likely to become a grandmother. While the Tangled Bank hypothesis might work in theory, it wasn't very likely. The different kinds of bodies built by the different sets of genes had to be quite distinct from each other in order for it to work. Nevertheless, it was the dominant idea at the time.

Lively found himself in New Zealand in 1985 because his wife, Lynda Delph, wanted to study evolutionary biology at the University of Canterbury. Lively got a job there as a postdoctoral researcher, and he wondered if New Zealand might offer him a way to test the different explanations for sex. In evolutionary biology, ideas tend to bubble up fast and easily, and often turn out to be miserably untestable. To test explanations for sex, Lively would have to find the right species to study. It would have to be a mix of sexuals and asexuals. Among some animal species, for instance, there are populations of males and females that live alongside clones. Other species are hermaphrodites, and they can choose to have sex with themselves or with another animal. Only in these sorts of animals could the generation-by-generation effects of evolution be seen, because a biologist could compare how the sexuals and asexuals fared. "If you're dealing with something that's all sexual," says Lively, "it's hard to know what selection would be for or against an asexual. But if you have a system where you have both, now you have the basis for comparison." He couldn't test an idea about the persistence of sex in humans, for instance, because we all do it. There is no lost tribe out there who can have children with natural cloning. In our own evolutionary lineage, the race between the sexuals and the asexuals ended hundreds of millions of years ago.

As luck would have it, there was a snail in New Zealand that fit Lively's research perfectly. Named *Potamopyrgus antipodarum*, the quarter-inch snail lived in most lakes, rivers, and streams in the country. While most populations of the snail were all identical clones, the product of parthenogenesis, some

were divided into male and female forms that used sex to reproduce.

Lively set out to see if the habitats of the snails had any influence on how they reproduced. The snails that lived in the
streams faced sudden floods, while the ones that lived in lakes
enjoyed a peaceful, stable existence. According to the Lottery
hypothesis, the snails in the streams should favor sex because
they had to survive in an unstable place. According to the Tangled Bank hypothesis, there would be more competition in the
lakes for different niches, and the males would be in demand
there.

Lively hiked to the high mountain lakes where the snails
lived and waded into the waters with his net. He gathered the
snails there, and to determine their sex, he cracked open their
shells and cut them open, looking for a penis behind their right
tentacle. But when he looked inside the snails, he was baffled—
they were packed with what looked to him like giant sperm. "I
showed them—unfortunately for me—to one of the parasitologists at the university, and he said, 'They're not sperm, you id-
iot, they're worms.'" The parasitologist explained to Lively that
the parasites were flukes that castrated their snail hosts, multiplied, and eventually got into their final host, a duck. In some
places, the parasitologist told him, the snails were riddled with
the flukes, and in others they were free of them.

The humiliation wasn't hard to handle, though, because
Lively realized that these parasites might let him test a third explanation for the endurance of sex: that parasites were responsible. The idea had been offered up in various forms by various
scientists, but most fully in 1980 by an Oxford University biologist named William Hamilton. Hamilton argued that when
hosts are faced with the Red Queen, sex can be a better strategy
for fighting parasites than cloning.

Consider a bunch of amoebae that reproduce by cloning
and that are divided up into ten genetically distinct strains. Let's
say that bacteria infect them and the Red Queen's race begins.

The bacteria come in strains of their own, each adapted to a different strain of host. The most common strain of amoebae are pounded down by their strain of bacteria, and when that strain of amoeba loses enough numbers, the parasitic spotlight switches to a different strain. Because these amoebae clone to reproduce themselves, every new generation of amoebae will be genetically identical with their forebears. The bacteria sweep through the same ten strains over and over again, and after a while, they may drive some of those strains into extinction.

Now imagine that some of these amoebae evolve the means to have sex. The males and females make copies of their genes and join them together to form their offspring's DNA, and as the genes combine, they get shuffled around. As a result, the offspring isn't a carbon copy of one of its parents but a new jumble of tier genes. Now the parasites have a much harder time chasing their hosts. Because the genes of the sexual amoebae mix, they no longer come in distinct strains, and it becomes harder for parasites to get a lock on them. The Red Queen still takes sexual organisms for an endless run, but their offspring may have less of a chance of getting infected. And the protection that this diversity brings to the sexual amoebae might give them a crucial edge in their competition with asexuals.

It was an elegant idea, but Lively didn't actually believe it when he first read about it. "My feeling—and I think it was general—was that it was a very clever idea, but it seemed unlikely to me to be true. The reason is that I just didn't see much parasitism in the world. If you're going to have a selective pressure that's intense enough, it should be something that has big, immediately obvious effects. At least in humans in this country, we don't see those big effects. And the people doing field biology were mainly interested in competition or predation. There was no tradition in parasites."

But the fact was that most animals—Lively's snails included—are rife with parasites. On the outside chance Hamilton might be right, Lively decided to start noting whether or not his snails were infested with the flukes. "The theory for parasites was just being laid by Hamilton in 1980, 1981, 1982, but no one had discovered systems where you could test them. I didn't know I was dealing with one until I started cracking open these snails. I realized it would be able to address Hamilton's idea, but if they had been viruses I would not have known it. Here we're dealing with big honking swimming worms, and anyone can see them under a dissecting microscope."

It didn't take Lively long to see a clear pattern. The snails in the lakes were more infected with the flukes than the ones in the streams, and it was in the lakes that there were the most males. The more infested a given lake was, the more males it held. The only hypothesis that could account for all three patterns was the Red Queen: in places where there were more parasites, there was a stronger evolutionary pressure for sex. "I was completely surprised. When I had half the data set I eventually published, I thought, 'Wow, there's a trend setting up.' So I went out and got a lot more data to see if it went away. It didn't. Adding more lakes didn't change it—it wasn't a few lakes that were highly sexual and highly infected."

Lively published those first results from New Zealand snails in 1987. He has made the study of sex his preoccupation ever since. He's tested the Red Queen hypothesis in other ways and found more support for it. In 1994, for example, he traveled to Lake Alexandrina on the southern island of New Zealand with his postdoctoral student Jukka Jokela. They gathered snails from both shallow and deep waters. In the shallow waters the snails live alongside ducks, which are the final hosts for the flukes, and the ducks shed the flukes' eggs there. With so many eggs in the water, the snails are sicker in the shallows than farther from the shore. Lively and Jokela found that there are more males among the snails in shallow water as well, probably as a result of the pressure of parasites. In a single lake, they could see parasites shaping the sex lives of their hosts.

At the same time, Lively has watched other biologists find

the Red Queen at work in other species. In Nigeria there lives another snail named *Bulinus truncatus*, one of the species that carry the blood flukes that cause schistosomiasis. Its sex life is more exotic than that of Lively's New Zealand snails. Every one is a hermaphrodite, with male and female gonads it can use to fertilize its own eggs and produce clones. But some of them also come equipped with a penis, which they can use to mate with other snails.

As with the New Zealand snails, it seems like a huge waste of effort for the Nigerian species to grow a penis and have sex when it can just fertilize itself. And as in New Zealand, parasites seem to make the effort worthwhile. According to the parasitologist Stephanie Schrag, each year the snails have a penis season. The waters are coolest in northern Nigeria in December and January. The snails use the cool temperature as a cue to produce more offspring equipped with penises—snails, in other words, that can mate with other snails. With more penises, there's more sex among the snails, and more shuffling of their DNA, and more variation in the next generation. The snails need about three months to mature, so this new sexually produced generation comes of age between March and June. And March to June happens to be the time of year when flukes are at their worst in northern Nigeria. In other words, snails seem to use sex to prepare months in advance for an annual attack of parasites.

The most unexpected support for the Red Queen's effect on sex has come from parasites themselves. Like their hosts, many parasites have sex, and in 1997, Scottish scientists asked why parasites bother. Like Lively, they looked for a species that isn't stuck reproducing only sexually or asexually. They chose *Strongyloides ratti*, a nematode that, as its name suggests, lives inside rats. The females living in the guts of rats lay eggs without any help from males. Once these eggs leave the rat's body they hatch, and their larvae emerge as one of two different forms.

One form is all female, and it spends its time looking for a rat to penetrate. It gets into the skin of the rat and then glides through it until it reaches the rat's nose. There it finds the nerve endings that the rat uses to smell, and it follows them into the brain. From there the parasite takes a route—no one knows the details—all the way to the rat's intestines, and starts making female clones again.

The other form of the nematode hatches from eggs in the soil and stays there. When the larvae mature, they turn into both females and males rather than females only, and instead of cloning they have sex to reproduce. The females lay fertilized eggs, giving birth to a new generation of worms that can penetrate the skin of rats and get back into their gut. In other words, *Strongyloides* can complete its life cycle with sex or without.

The Scottish scientists decided to see whether a change in the immune system of a rat might influence the kind of reproduction the parasites chose. They put *Strongyloides* into rats, and the rats mounted an immune response to the parasites. They then gave the rats shots of antiworm medicine to clear the parasites out of their bodies. Now the rats were primed to fight off a second invasion. When the scientists reinfected the rats and the new wave of nematodes began making eggs, the parasites that emerged from them were more likely to be sexual forms. In another experiment, the scientists depressed the immune system of a rat with radiation and then infected it with *Strongyloides*. They found that the parasites were much more likely to clone themselves than to have sex.

These experiments showed that *Strongyloides* would prefer to reproduce asexually, but a healthy immune system forces it to have sex. "Your immune system is a sort of parasite of the parasite," says Lively. Like parasites, T cells and B cells multiply into many different lineages, and the most successful killers get to reproduce themselves the most. Like their hosts, parasites can defend themselves by having sex and diversifying their genes.

All the work that Lively and these other scientists have done on the origins of sex rests on the Red Queen's shoulders, and yet it has been hard to get a glimpse of the Queen herself. Some researchers who run computer simulations of the struggle between host and parasite have seen her shadow flit across their monitors. In Lively's own work, he could see her effects only by mapping where the sexual and asexual snails lived—taking a snapshot of her effects at a given instant. But eventually he had studied enough snails to see her work spread across time rather than space.

For five years he and another of his postdoctoral students; Mark Dybdahl, netted snails in Lake Poerua. The snails there were all clones, and most of them belonged to four main lineages. Lively and Dybdahl took a census of the four snail clans each year and watched their populations rise and fall. They took the rarest clones and the common ones to their lab at Indiana University, where Lively now works. There they exposed both kinds of snails to their flukes. They found a huge difference: the parasites had a much harder time infecting the rare snails than the common snails. Here was a central prediction of the Red Queen: that being rare gives an organism an advantage, because parasites are more adapted to the more common hosts.

They then looked at their census of the snails of Lake Poerua over five years. In a given year, they found, there wasn't much of a connection between the number of parasites infecting a lineage of snails and how big the lineage was. The ones with heavy burdens of parasites weren't the most common. But with a five-year record, Lively and Dybdahl could look back at the lineages in previous years. When they did, a distinct pattern jumped out. The snail lineages that carried the heaviest burdens of parasites in a given year had been the most common snails a few years before, and now they were declining. The snails had started out rare and had increased their numbers, but eventually the parasites caught up with them and started driving their numbers down. Because it took a while for the evolution of the flukes to catch up with their hosts, the flukes reached their greatest success only after the snails had already started to decline.

For the first time, scientists have been able to see the Red Queen at work, by moving back through time. It's a method that Alice would have approved of. At one point in her adventures, she lost sight of the Red Queen. She asked the Rose how to catch her, and the Rose replied, "I should advise you to walk the other way."

"This sounded like nonsense to Alice, but after only occasional glimpses of the Queen in the distance, she thought she would try the plan, this time, of walking in the opposite direction. It succeeded beautifully. She had not been walking a minute before she found herself face to face with the Red Queen."

• • •

Shortly after William Hamilton proposed that parasites drive the evolution of sex, he realized that this idea naturally gave rise to another one. Sex may help organisms fend off parasites, but it brings trouble of its own. Say you're a hen, and your genes are particularly well suited to fighting off the parasites that the Red Queen has made most common at the moment. You want to have some chicks, but to do that you have to find a rooster, and half of the chicks' genes will have to come from him. If you pick a rooster that has bad parasite-fighting genes, your chicks will suffer the consquences. It pays for you to be picky about your mates and to try to figure out which roosters have good genes. The rooster doesn't have to be as picky, because he can make millions of sperm. You, on the other hand, can raise only a few dozen eggs over your lifetime.

Working with a graduate student, Marlene Zuk, at the University of Michigan, Hamilton suggested that females judge male displays to decide how well they can fight parasites. A weak suitor will have to spend most of his efforts fighting off parasites and will have very little resources left over. But a male who can resist parasites will still have enough energy left over to advertise his healthy genes to females. These advertisements, Hamilton and Zuk argued, should be showy, extravagant, and expensive. A rooster's comb might qualify as just this sort of biological resumé. It serves no particular purpose in the rooster's survival. In fact, it's a burden to him, because in order to keep it red and puffy, the rooster has to pump testosterone into it. Testosterone tends to depress the immune system, putting roosters at a disadvantage in fighting off parasites.

Just as parasites might create the rooster's comb, they might draw out the long tail feathers on birds of paradise. They might make redwing blackbirds redder, they might put bright spots on male stickleback fish, and they might make the sperm packages of crickets bigger. Anything that females could use to judge males might be influenced by parasites.

Hamilton and Zuk presented their idea in the early 1980s, with a simple test. You'd expect that on the whole, the members of a species saddled by many parasites would be showier than a species with a lighter load. According to their hypothesis, bacteria and viruses wouldn't have a big impact on male display. They tend to kill their hosts or get killed by them. In the first case, there's no male left to do the displaying; in the second, a sick male could recover so well he'd be indistinguishable from stronger males.

Hamilton and Zuk gathered reports on North American songbirds and their parasites that cause chronic, grinding diseases—bird malaria, for example, and *Toxoplasma*, trypanosomes, and various worms and flukes. They then rated the showiness of the males of each species in terms of their brightness and their song, and found that the species with the most parasites had the strongest male displays.

That initial work inspired a huge amount of research (more, actually, than Hamilton's broader theory on the origin of sex itself). Zoologists tested these ideas in the songs of crickets, in the spots on stickleback fish, in the throat pouches of fence lizards. In many of the tests—especially the lab experiments—Hamilton and Zuk fared well. Zuk studied red jungle fowl from Southeast Asia, for example, which are wild relatives of chickens. She kept track of the choices made by female jungle fowl in her lab and measured the combs on the males they chose. Females, she found, consistently preferred males with longer combs.

In a more elaborate study, Swedish scientists studied wild ring-necked pheasants. Male pheasants have spurs on their legs, and the researchers found that the females used the length of the spur to decide which male to mate with. The researchers then looked at the immune system genes of the pheasants and found that the pheasants with the longest spurs shared a particular combination of genes. They don't know what those genes actually do to help the males fight off parasites. But they observed the offspring of the pheasants and found that the ones with long-spurred fathers had better chances of surviving than those with short-spurred ones.

There's no reason why these antiparasite advertisements can't extend beyond a male's body to the way he courts females. That certainly seems to be what's going on with the fish *Copadichromis eucinostomus*, which lives in Lake Malawi in central Africa. To attract females, the males build bowers out of sand on the lake bottom. Some of them are nothing more than a handful of grains sitting on top of boulders, while others are big cones several inches high. The males build their bowers together, creating dense neighborhoods, and each defends his own against roaming males that are trying to usurp him. The female fish spend most of their time feeding on their own, but when the time comes to mate, they go to the bower neighborhood and inspect the males' work. If a female chooses to mate with a male, she releases an egg and puts it in her mouth. The male puts his sperm in her mouth and she carries away the fertilized egg.

The females apparently use the bowers to find out which males do the best job of fighting parasites such as tapeworms. Experiments have shown that the females prefer males who built big, smoothly shaped bowers, and these males also happen to be the ones who carry the fewest tapeworms. A fish that's carrying tapeworms may have to spend so much time eating that it can't maintain its bower. The bower thus becomes a medical chart, and perhaps a genetic profile.

But the Hamilton-Zuk hypothesis has failed several tests as well. Male desert toads attract their mates with their calls, for example, but a loud call doesn't reflect an immune system better able to fight off *Pseudodiplorchis*, the parasite that lives in their bladder and drinks their blood. In some species of fence lizards, the males have brightly colored throat flaps that females just adore, but there's no connection between their brightness and parasites such as *Plasmodium* that attack the lizards.

These failures have made scientists wonder whether they've been testing the Hamilton-Zuk hypothesis the wrong way. A particular parasite may be harmful or harmless, and may therefore have a big influence on a male's display or none at all. If you have a lot of studies on the loads of different parasites, it's hard to use them to come up with any sort of general conclusion. Rather than counting the parasites themselves, measuring the immune system may be more reliable. Immune systems have evolved to cope with many different kinds of parasites, so they can offer a better overall clue. It's a lot harder to count microscopic white blood cells than giant tapeworms, but it turns out to be a better method. Immune studies give the Hamilton-Zuk hypothesis strong, consistent support. Peahens, for example, choose peacocks with more extravagant tails, and researchers have found that peacocks with more extravagant tails have immune systems that can mount a stronger response to parasites.

Another reason why the Hamilton-Zuk hypothesis is falling short may be that scientists are looking at the wrong signals.

They've stuck with visible cues like rooster combs and lizard pouches because they're easy to measure. But among the channels of communication between the sexes, vision may not be all that important. Mice, for example, can smell the urine of a prospective mate and tell whether or not it's carrying parasites; if a male mouse is sick, a female will stay away. It's even possible that males could use their odors to advertise their strength against parasites with some kind of extravagant, irresistible perfume. "The scent of a male mouse," writes one biologist, "is the chemical equivalent of a peacock's plumage."

And even if Hamilton and Zuk's idea turns out to fail for other animals, parasites may well have shaped their sex lives anyway for very different reasons. Once again, it all comes down to how a given animal passes on its genes. Among bees, young queens leave their birthplace hive at the end of the summer with a retinue of males. After she mates with them, the males then die, but the queen survives the winter and emerges in the spring to start a new colony with the eggs that were fertilized the previous fall. Every species of bees, in other words, flows through the bottleneck of its few queens.

By studying the DNA of bees, biologists have found that the queens may mate with ten or twenty males during their nuptial flight. That much sex, pleasure aside, is costly: a mating queen is more vulnerable to a predator's attack, and she could save the energy involved in all that sex to survive the winter.

Bees may be having all that sex as a defense against parasites, as demonstrated by Paul Schmid-Hempel, a Swiss biologist. He injected sperm into queens and then raised the colonies the queens gave birth to. Some queens got the sperm of only a few closely related males, while others got a cocktail with four times more genetic diversity. When the queen's colony began to hatch, Schmid-Hempel put the colonies out in a flowering meadow near Basel and left them there until the end of the season, when he went out to capture them.

By just about every measure, the offspring of high-diversity queens were far stronger against parasites than were the low-diversity ones. Their colonies had many fewer infections, fewer kinds of parasites invading, and fewer parasites in a given individual. The offspring of high-diversity queens were more likely to survive till the end of the summer, which made it more likely that they'd produce future colonies. Instead of carefully eyeing up a single male to mate, a queen bee may look for many suitors to create a genetic rainbow in her future hive.

●　　●　　●

As critical as an immune system may be to surviving parasites—particularly an immune system that can evolve rapidly—it's really a defense of last resort. It fights against invaders that have already crossed the moat and are inside the castle. It would be far better to keep the parasites from getting in at all. Evolution has obliged. Hosts have adapted to fight off parasites with the shapes of their bodies, their behavior, the way they mate, even the shape of their societies—all designed to keep parasites at a distance.

Many insects are shaped expressly to fend off parasites. During their larval youth, some species are covered in spikes and tough coats that discourage wasps from trying to lay their eggs inside. Some have tufts of detachable barbs on their bodies, which entangle a wasp when it tries to land on them. When butterflies form cocoons, they sometimes dangle them from a long thread of silk that makes it impossible for wasps to get enough leverage to stab through their coat.

For some insects, armor is not enough. Thousands of species of ants, for example, are tormented by thousands of corresponding species of parasitic flies. The fly perches above the trail made by the ants from their nest to their food. When a suitable ant passes underneath, the fly dives down onto the ant's back and wedges its egg-laying tube into the chink between the ant's head and the rest of its body. Quickly the eggs hatch, and the maggots chew their way into the ant's interior and then travel to the ant's head. These larvae are muscle eaters. In a mammal it might make sense for them to make their way into a bicep or a thigh, but in ants the fleshiest place is the head. Unlike our brain-crammed skulls, those of ants hold only a loose tangle of neurons, the rest of the space being dedicated to muscles that power its biting mandibles. A maggot inside an ant's head chews on the muscles, carefully avoiding the nerves, and grows until it fills the entire space. Finally, one day, the ant meets its awful end: the parasite dissolves the connection between the head and the rest of the body. Like a ripe orange, it drops to the ground. While the headless host stumbles around, the fly begins its next stage, forming its pupa. Other insects have to weave their cocoons exposed to the elements and hungry predators, but the fly develops snug in the tough cradle of an ant's head.

These flies are so destructive that ants have evolved defensive manuevers against them. Some will run to escape the flies; others stop in their tracks and begin flailing wildly, gnashing their mandibles as soon as they even sense that a fly is overhead. A single parasitic fly can stop a hundred ants in their tracks along six feet of their trail. If the fly lands on the back of one species and gets ready to lay its eggs behind the head, the ant suddenly snaps its head back against its body, crushing the fly in its vise.

Among the leaf-cutting ants, these flies have transformed their entire social structure. Leaf-cutting ants travel from their nests to trees, hack off foliage, and take it back home, forming a parade of green confetti on the forest floor. Leaf-cutters are the dominant herbivores in many forests of Latin America—wildebeest in miniature, although they don't actually eat the leaves. Instead, they bring them home to their colonies, where they use them to grow gardens of fungi, which then become their meal. If you want to get technical, leaf-cutters aren't so much herbivores as mushroom farmers.

Leaf-cutter colonies are divided into big ants, which carry

the leaves home, and little ants. The little ants (known as minims) tend the gardens, and they can be also found riding atop the leaves being brought home by the big ants. Entomologists have puzzled for a long time over why the minims would waste their time hitching rides like this. Some suggested that they must collect some other kind of food on the trees, maybe sap, and then go home on the leaves in order to save energy. In fact, minims are parasite guards. The parasitic flies that attack leaf-cutters have a special approach to their hosts: they land on the leaf fragments and crawl down to where the ants grip it in their mandibles. The fly then lays eggs in the gap between the mandible and the ant's head. The hitchhiking minims patrol the leaves or perch on top, their mandibles open. If they encounter a fly, they scare it away or even kill it.

For bigger animals, the struggle with parasites is just as intense, although it's not as obvious as an ant wrestling a fly. Mammals are continually assaulted by parasites—by lice, fleas, ticks, botflies, screwworms, and warbleflies—that suck blood or lay their eggs in the skin. In response, mammals have evolved into obsessive groomers. The way a gazelle lazily flicks its tail and nuzzles its flank may look like the picture of peace, but it's actually in a slow-motion struggle against an army of invaders. The gazelle's teeth are shaped like rakes, not to help it eat but to scrape away lice and ticks and fleas. If its teeth are blocked, its load of ticks will explode eightfold. Gazelles don't groom themselves in response to any particular scratch; they clean themselves according to a clocklike schedule because parasites are so relentless. Grooming cuts into the time an animal needs to eat and guard against attacks from predators. The top impala in a herd ends up riddled with ticks—six times more than females—because he is too busy staying vigilant against male challengers.

The shape of an animal's society may also help cut down on parasites. Animals protect themselves from predators in this way. Fish that stay in schools can pool their vigilance; as soon as any of them senses a predator, they can all swim away. And even if the predator should attack, each member of the school has lower odds of being killed than if it were on its own. It's time to put the parasite alongside the lion. Increasing the size of a herd not only will lower the odds that each gazelle will be eaten by a lion, but also will lower the odds that each individual will be attacked by a tick or some other blood-sucker. On the other hand, parasites may simultaneously keep herds from getting too big. As animals crowd together in bigger and bigger groups, they make it easier for some parasites to be passed from host to host, whether they are viruses carried on a sneeze, fleas passed on with a nuzzle, or *Plasmodium* carried by a hungry mosquito.

Parasites may even teach animals manners, according to Katherine Milton, a primatologist at the University of California at Berkeley. Milton studies the howler monkeys of Central America, and she's been struck by the viciousness of one of their parasites: the primary screwworm. This fly searches for open wounds on mammals; it can even find the hole made by a tick bite. It lays its eggs inside the wound, and the larvae that hatch start devouring their host's flesh. They do so much damage in the process that they can easily kill a howler monkey.

The screwworm may make howler monkeys leery of fighting with each other over mates or territory. The fight might only be a minor scuffle, but if a monkey gets a scratch, a screwworm could make it the last scuffle it ever has. Screwworms are so efficient at finding wounds, in fact, that evolution may frown on violent howler monkeys. Instead, it may have made them affable creatures, and it may have encouraged them to evolve ways to confront each other without getting hurt, such as howling and slapping rather than biting and scratching. There are many other mammals that also have ways to avoid fights, and it's possible that they are also trying to avoid parasites.

The best strategy for a host is simply not to cross paths with a parasite at all. Some of the adaptations hosts make to avoid the notice of parasites are so grotesque, so outrageous, that it's hard to tell at first that they actually are designed for parasites at all. Consider leaf-rolling caterpillars. They're pretty ordinary insect larvae with one exception: they fire their droppings like howitzers. As a bit of frass starts to emerge from the caterpillar it pushes a hinged plate back against a ring of blood vessels surrounding its anus. The blood pressure builds up behind the plate, which the caterpillar then releases. The pressure of the blood slams against the droppings so suddenly that it blasts them three feet a second, in a soaring arc that carries them up to two feet away.

What on Earth could have driven the evolution of an anal cannon? Parasites could. When parasitic wasps home in on a larva such as the leaf-roller caterpillar, one of the best clues is the odor of their host's droppings. Since caterpillars are sedentary, not racing from branch to branch, their droppings will normally accumulate close by them. The intense pressure put on leaf-roller caterpillars by wasps has pushed the evolution of high-pressure fecal firing. By getting their droppings away from them, the caterpillars have a better chance of not being found by wasps.

Vertebrates, like insects, will also go out of their way to avoid parasites. Cow manure fertilizes the grass around it, making it grow lush and tall, but the cows generally stay away. They keep their distance because the manure often carries the eggs of parasites such as lungworms, and the parasites that hatch from them crawl up the neighboring blades of grass in the hope of being eaten by a cow. Some researchers have suggested that mammals that make long migrations, such as caribou and wildebeest, plot their course in part to avoid parasite-thick spots along the way. Swallows will fly back to their old nests and reuse them, unless they discover that their nests have been infested with worms and fleas and other parasites, in which case they'll build a new one. If baboons discover that the area where they sleep has been overrun with nematodes, they'll go away and won't return until the parasites have died away. Purple martins go so far as to line their nests with plants like wild carrot and fleabane that contain natural parasite-killers. Owls sometimes catch blind snakes, but rather than tear them apart to feed their chicks, they drop them into their nests. There the snakes act as maids, slinking into the nooks of the nest and eating the parasites they find there.

•   •   •

Even if your mother was an excellent judge of fish bowers, even if you perfected your fly-killing head-snap, even if you can blast your frass into the neighboring meadow, you may end up with a parasite inside you. Your immune system will do its level best to stave off the invasion; it's an exquisitely precise system of defense brought about thanks to the evolutionary pressure of parasites. But hosts have evolved other kinds of warfare. They can enlist other species to help them; they can medicate themselves; they can even reprogram their unborn offspring to prepare for a parasite-ridden world.

When a plant is attacked by a parasite, it defends itself with its own version of an immune system by creating poisonous chemicals that the parasite eats as it chews on the plant. But it also fights by sending out cries for help. When a caterpillar bites a leaf, the plant can sense it—a feeling not carried by nerves but felt nevertheless. And in response, the plant makes a particular kind of molecule that wafts into the air. The odor is like perfume for parasitic wasps; as they fly around searching for a host they are powerfully lured by the plant's smell. They follow it to the wounded leaf and find the caterpillar there, and they inject it with eggs. These conversations between plants and wasps are not only timely but precise. Somehow the plant can sense exactly which species of caterpillar is dining on it and spray the appropriate molecule into the air. A parasitic wasp will respond only if the plant lets it know that its own species of host sits on a leaf.

Animals will sometimes defend themselves against parasites with a change of diet. Some will just stop eating—if a sheep is hit by a bad dose of intestinal worms, for instance, it may graze only a third of its normal intake. Such a change clearly can't benefit the parasite, which wants the sheep to eat a lot so that it can eat a lot and make a lot of eggs. Researchers suspect that eating less may somehow boost the host's immune system, making it better able to fight the parasite. On the other hand, the animals may not be simply fasting but may be being choosier about what they eat, choosing food that has the right nutrients to help them fight the infection.

Sometimes animals under attack by parasites will start eating foods they almost never eat. Some species of woolly bears, for example, normally eat lupine. They sometimes get attacked by parasitic flies that lay eggs in their bodies. Unlike the flies that attack ants or other insects, though, these parasites don't always kill their hosts when they emerge from their bodies. And the woolly bears improve their own odds of survival by switching from a diet of lupine to one of poison hemlock. The parasitic flies still crawl out of their bodies, but some chemical in the hemlock helps the woolly bears stay alive and grow to adulthood. The woolly bears, in other words, have evolved a simple kind of medicine. Medicine may be pretty widespread among animals—there are plenty of records of animals sometimes eating plants that can kill parasites or expel them out of their gut. But researchers are still trying to prove that they actually eat those foods when they get sick.

When things get truly bleak—when there's little hope a host can kill a parasite inside it—it cuts its losses. It has to accept that its life is doomed. Evolution has given hosts ways to make the best of the time they have left. When some species of snails are infected with flukes, there's only a month or so before the parasites castrate them and turn them into nothing more than food-gathering slaves. That still gives the snails a month to produce the last of their offspring. They take full advantage, producing a final burst of eggs. If a fluke gets into a snail that's still sexually immature, it will respond by developing its gonads much faster than if it were healthy. If they're lucky, the snails can squeeze out a few eggs before the parasites cut them off.

When the fruit flies of the Sonoran desert are attacked by parasites, their response is to get horny. They feed on the rotting flesh of the saguaro cactus, and sometimes they encounter mites there. The mites leap onto the flies and jab their needle-like mouths into their bodies, sucking out their internal fluids. The consequences can be grave—a heavy infestation of mites can kill a fly in a few days. Biologists have found a big difference between the sexual activities of healthy and mite-infested male fruit flies. The parasites trigger the males to spend more time courting females, and the more parasites a male has, the more time he spends doing so, in some cases tripling his efforts.

At first this might seem like another display of puppetmastery, as a parasite speeds up its own transmission by putting infected flies in contact with healthy ones. In fact, the mites seem to get on flies only when they feed on cactus. They never hop from one mate to another. It appears that parasites have essentially driven flies to evolve a habit of mating more when death—and no more matings—seems imminent.

Why don't the flies make the fast-and-furious lovemaking style a permanent one? The answer, probably, is that the mites aren't always assaulting the flies. Some cactuses are covered with them; others are mite-free. As with bees, sex puts a lot of demands on fruit flies, making them an easy target for predators. Better to be flexible, mating at a slower speed normally and speeding up in the face of parasites.

Lizards are also tormented by mites of their own; they can die from an infestation, and the survivors are likely to have their growth stunted. But when they're attacked, they go through a different sort of change: they alter their unborn offspring. A lizard infested by mites produces babies that are bigger and faster than those born of healthy parents. A healthy baby lizard will have a growth spurt in its first year and then grow more slowly for the rest of its life. But a lizard born to mite-ridden parents will grow fast for its first two years or more. Lizard mothers apparently can program the growth of their offspring to adapt to the presence of parasites. If there are no mites around, their offspring can grow slowly and live a long life. But if mites turn up, it pays to grow faster in order to reach a healthy weight as an adult, even if that means dying sooner.

And if a host is doomed to die, it can do its best to spare its kin. Worker bumblebees spend their days flying from flower to flower, collecting nectar and bringing it back to their hive. At night they stay in the hive, kept warm by the heat of thousands of flapping wing muscles. On its travels for nectar, a bumblebee may be attacked by a parasitic fly, which lays an egg in its body. The parasite matures within the bumblebee, and in the warmth of a beehive its metabolism runs so quickly that it can finish growing up in only ten days. The fly emerges from its host and can infect the rest of the hive. Yet, many parasitic flies don't get that luxury because their host does something strange: it starts spending its nights outside the hive. By staying out in the cold, the worker slows down the parasite's development. It also prolongs its own life. The combined effect makes it unlikely that the parasite will ever make it to maturity before the bee itself dies. In this way, the bumblebee prevents an epidemic from breaking out in its hive.

As cunning as these kinds of counterattacks may be, parasites can evolve counter-counterattacks. If a cow avoids manure to keep away from the lungworms it holds, the parasites will leave the manure. When a lungworm drops to the ground in the manure, it bides its time until light strikes it. That is its signal to climb upward until it reaches the surface of the manure. It begins to hunt around for a species of fungus that is also a parasite of cows—a species that also responds to light by growing little spring-loaded packages of spores. As soon as the lungworm touches the spore package, it latches on and climbs up to the top. The fungus catapults itself six feet into the air and soars away from the manure. The lungworm rides it like a puddle-jumper, and out of range of the manure it has better odds of being eaten by a cow.

Study arms races long enough, and you start to imagine that hosts and parasites could carry each other into the clouds, each driving the evolution of its counterpart so hard that they become all-powerful demigods hurling lightning bolts at each other. But of course the race has limits. When Kraaijveld set his wasps against his fruit flies, the fruit flies reached a 60 percent resistance to the wasps after only five generations, but in later generations the resistance simply stayed there at 60 percent. Why didn't it keep rising to 100 percent, creating a race of perfectly immune flies? Fighting parasites comes at a high cost. It requires energy to make the necessary proteins—energy that can't be channeled somewhere else. Kraaijveld set his flies selected for wasp-fighting in competition against regular flies for food and found that they fared badly. They grew more slowly than the flies that were still vulnerable to the wasps, they died young more often, and when they grew into adults they were smaller. Evolution doesn't have an infinite arsenal to offer hosts, and at some point they have to relent, to accept that parasites are a fact of life.

●　　●　　●

When Darwin set out to write *The Origin of Species*, figuring out how natural selection works wasn't his ultimate goal. It was really only a means to an end—to explain the title of his book. After branching and growing for 4 billion years, the tree of life today wears a heavy crown. Scientists have found 1.6 million species, and they may be only a sliver of Earth's full diversity, which may be many times larger. Darwin wanted to know how that diversity came to be, but he didn't know enough about biology to find the answer. Now that scientists have a better understanding of heredity and how genes rise and fall over the generations, they're closing in on how species actually come into being. And they're finding that the race between hosts and parasites is crucial once again. It may account for much of life's dense evolutionary canopy.

A new species is born out of isolation. A glacier may cut off a pocket of mice from the rest of their species, and over the course of thousands of years they may develop mutations that make them unlike the rest of the mice and unable to mate with them. A single species of fish may come into a lake and some of its members may start specializing in feeding on the mucky bottom, others in the clear shallows. As they evolve equipment for each kind of life, crossbreeds will turn out badly suited to either one. Natural selection will push them apart, and they will stay more and more with their own until they form separate species.

The life of a parasite encourages new species to form. Parasites can adapt to a single nook in a host—a curl of the intestines, the heart, the brain. A dozen parasites can specialize on the gill of a fish and subdivide it so precisely that there's no competition between them. Specializing on particular host species makes parasites even more diverse. A coyote will eat just about anything on four legs, and partly as a result, there is only one species of coyote in all of North America. Unlike coyotes and other predators, many parasites are under the Red Queen's control. A parasite that prefers many different hosts has to try to play the Red Queen game with all of them, like a chess player running frantically between a dozen games he's playing at once. If another parasite should undergo a mutation that makes it prefer only one host, all of its evolutionary effort will be focused on that host alone. The hosts don't even have to be an entire species—if just a population of the host is isolated enough, it will pay for the parasite to specialize only on them. With parasites focusing so much on a species or a fraction of a species, they leave room for other parasites to evolve.

As new species are born older ones are going extinct. Species disappear when they are outcompeted, when their numbers shrink down below a critical threshold, or when the world changes too quickly for them to adapt. Lineages of parasites may be able to resist extinction better than those of free-living creatures. While parasites tend to be specialists, they also dabble a little from time to time. Sometimes a new host will turn out to be a good home, and the parasite may found a new species. Tetrabothriid tapeworms are still with us, living in puffins and gray whales, for example, but the pterosaurs and ichthyosaurs in which they lived 70 million years ago are not. The diversity of parasites is like a great lake, with big streams of new species flowing in but only a trickle flowing out into extinction.

Take all these reasons together, and it's not so surprising that there are so many species of parasites. There are about four thousand species of mammals, and aside from a few rabbits and deer waiting in some obscure forest to be discovered, that number is firm. But there are five thousand species of tapeworms known so far, and new species are discovered every year. There are two hundred thousand species of parasitic wasps. The insects that are parasites of plants number in the hundreds of thousands as well. Add them all up, and the majority of animals are parasitic. Untold thousands of fungi, plants, protozoa, and bacteria also proudly bear the title of parasites.

It's now becoming clear that parasites may have pushed their hosts to become more diverse as well. Parasites don't attack an entire species in the same way. The parasites in a particular region can specialize on that population of hosts, adapting to that local set of host genes. The hosts evolve in response—but only the hosts in that region, not the species as a whole. This local struggle has produced some of the fastest cases of evolution ever documented—whether they be yucca moths and the flowers where they lay their eggs, snails and their flukes, or flax and their fungi. And as these populations of hosts fight off their dedicated parasites, they become genetically distinct from the rest of their species.

But this is actually only one way of many that parasites may be able to help turn their hosts into new species. Genetic parasites can speed up the evolution of their hosts, for instance. In order for evolution to take place, genes have to take on new sequences. That can happen with ordinary mutations—the occasional cosmic ray from outer space slamming into DNA or the sloppy crossing of genes as cells divide. But it can happen faster with the help of a genetic parasite. As it hops from chromosome to chromosome within a cell, or as it leaps from species to species, it can wedge itself into the middle of a new gene. This sort of rude arrival usually causes trouble, in the same way throwing a random string of commands into the middle of a computer program does. But every now and then, the disruption turns out to be a good thing, evolutionarily speaking. An interrupted gene may suddenly become able to make a new kind of protein that does a new sort of job. The blind jump of one genetic parasite seems to have made us able to fight parasites more effectively. The genes that make the receptors on T and B cells show signs of having been created out of the blue by genetic parasites.

And once a genetic parasite has established itself in a new host, it can disrupt the unity of the entire species. The typical fate of a genetic parasite is to explode through its host's genome during the succeeding generations, wedging itself into thousands of sites. As time passes, the hosts that carry it will diverge on their own into separate populations—not distinct species, but groups that tend to breed among themselves. As they do, the genetic parasite continues to hop from place to place in their DNA. Its hopping will be different in each population, and it will make their genes more and more different from one another. Eventually, when a Romeo and Juliet from the two populations meet and try to mate, their distinct collections of genetic parasites may make them incompatible. By making it harder for different populations of their hosts to mix their genes, the genetic parasites encourage them to split into new species.

Another way parasites might be able to create a new species

is by mucking up the sex lives of their hosts. A bacterium called *Wolbachia* lives in 15 percent of all insects on Earth as well as many other invertebrates. It lives within its host's cells, and the only way it can infect a new host is by colonizing a female's eggs. When the egg that *Wolbachia* lives inside becomes fertilized and grows into an adult, it grows up with a case of *Wolbachia* infection.

There's a downside to this way of life: if *Wolbachia* should grow up in a male it faces a dead end, because there are no eggs for it to infect. As a result, *Wolbachia* has taken control of its hosts' sex lives. In many of its host species, it tampers with the sperm of infected males so that they can successfully mate only with *Wolbachia*-carrying females. If one of these infected males should try to mate with a healthy female, all of their offspring will die. *Wolbachia* uses a different strategy in some species of wasps: normally these insects are born as males and females, which reproduce sexually, but when *Wolbachia* infects them, the wasps become female-only, able to mother only more females. By turning its hosts all female, the bacteria gives itself that many more hosts.

In both these cases, *Wolbachia* genetically isolates the infected hosts from the uninfected ones. A newly born host will be the offspring of either *Wolbachia*-carrying parents or two healthy ones. It won't be a healthy-unhealthy hybrid. By setting up this reproductive wall, the parasite may be able to set the stage for a new species to form. *Wolbachia* is only the best-known parasite out of many that tamper with their hosts' sex lives, so this may turn out to be a common way new species form.

Darwin always had a sharp sense of irony, but this one might have been too much for him to bear. To understand how life changes its form, how evolution is driven forward, and how new species come to be, he could have found inspiration in his dying children. When it comes to the tapestry of life, parasites are a hand at the loom.

7

The Two-Legged Host

Humanity has but three great enemies: fever, famine and war; of these by far the greatest, by far the most terrible, is fever.

—William Osler

The beauty of parasites is an inhuman one. It's inhuman not because parasites have come from another planet to enslave us but because they have been on this planet so much longer than we have. I sometimes think about Justin Kalesto, the Sudanese boy who was so racked by sleeping sickness that he could only whimper in his bed. He was twelve years old, and on his own he'd be no match against a dynasty of parasites that have lived in almost every sort of mammal—in reptiles, birds, dinosaurs, amphibians—everything backboned since fish came ashore, that have lived inside fish before anything walked on land, that have evolved their way into the guts of insects as well as vertebrates, that even thrive inside trees. The entire human race is a child like Justin: a young species perhaps only a few hundred thousand years old, a tender new host for trypanosomes and other parasites to make their own.

Of course parasites have never encountered a host quite like us. We can fight against them with inventions such as medicines and sewers as no animal has before. And we've changed the planet around us as well. After billions of years of glorious success, parasites now must live in the world we've made: a world of shrinking forests and swelling shanty towns, of vanishing snow leopards and multiplying chickens. But thanks to their adaptability, they're doing well overall. We should worry about the disappearance of condors and lemurs; their extinction will show us how badly we're stewarding the planet. But we shouldn't worry about the extinction of parasites. The tick species that live on black rhinos will probably disappear with their hosts in the next century. But there is no danger of parasites in general disappearing from the planet during the lifetime of our species; just about all of them will probably still be here when we're gone.

While parasites must live in the world we've made, the opposite is true as well. They have structured the ecosystems that we depend on, and they have sculpted the genes of their hosts for billions of years, our own included.

It is surprising just how precisely they've shaped us. When immunologists began studying antibodies, they found that they could sort them into categories. Some had hinged branches; some were built like five-rayed stars. Each group of antibodies has evolved to work against particular sorts of parasites. Immunoglobulin A works against the influenza virus and little else. The star-shaped immunoglobulin M staples its rays to bacteria like *Streptococcus* and *Staphylococcus*.

And then there was a strange little antibody called immunoglobulin E (IgE). When scientists first found this antibody, they couldn't figure out what it was for. It would remain at barely detectable levels in most people, except during a bout of hay fever or asthma or some other allergic reaction, when it would suddenly surge through the body. Immunologists have worked out how IgE helps trigger these reactions. When certain harmless substances get into the body—ragweed pollen, for example, or cat dander, or cotton fibers—B cells make IgE antibodies tailored to their shape. These antibodies then are anchored to special immune cells called mast cells that are found in the skin, the lungs, and the gut. Later, the harmless substance for which the IgE was made enters the body again. If it latches onto a single IgE antibody on a mast cell, nothing happens. But if it should latch onto two of them sitting side by side on the mast cell, the harmless substance switches it into action. Suddenly the the mast cell blasts out a flood of chemicals that make muscles contract, fluids pour in, and other immune cells flood the site. Hence the sneezing of hay fever, the wheezing of asthma, the red hives of a bee sting.

Since allergies serve no good purpose, immunologists could only look on IgE as one of the rare shortcomings of the immune system. But then they discovered that IgE can be good for something: fighting parasitic animals. IgE may be rare in the United States and the few other parts of the world that are now free of intestinal worms, blood flukes, and their like, but the rest of humanity (not to mention the rest of Mammalia) carry a heavy load of flukes, worms, and IgE. Experiments on rats and mice have shown that IgE is crucial for fighting these parasites; if animals are robbed of their IgE, they're overrun by parasites.

The immune system has, in a sense, recognized that parasitic animals are different from the other creatures that live in our bodies; they're bigger and their coats are far more complex than those of single-celled organisms. As a result, it has devised a new strategy against them that depends on the IgE antibody. Exactly how that strategy works isn't completely clear, and it may be a bit different for each parasite. It's been worked out best for *Trichinella*, the parasitic worm that grows up in muscle cells and then enters a new host in a piece of meat tumbling into the stomach.

Once *Trichinella* has thrashed its way free, it moves through its host's gut by spearing through the projections that line the bowels. Immune cells in the lining of the intestines pick up some of the proteins from the parasite's coat and travel to the lymph node that lies just behind the intestines. They present the *Trichinella* proteins to T cells and B cells in the node, setting off the creation of millions of cells targeting the parasite. These B and T cells then come pouring out of the lymph node and swarm through the lining of the intestines.

The B cells make antibodies, including IgE, which spread over the surface of the intestines and form a shield that *Trichinella* can't penetrate to anchor itself. At the same time, the mast cells are switched on, bringing on sudden spasms and floods through the intestines. Unable to get any purchase on the intestines, the parasites are washed away.

This precise strategy against a particular parasite—and many others—was in place long before our first primate ancestors swung through the trees 60 million years ago. And if monkeys and apes are any guide today, they needed all the help they could get: primates today are rife with parasites—malaria in their blood, tapeworms and other creatures in their intestines, fleas and ticks in their fur, botflies under their skin, and flukes in their veins.

At some point before 5 million years ago, our own ancestors, living somewhere in Africa, split off from those of today's chimps. Hominids began standing on two legs and gradually moving from lush jungles to sparser forests and savannas, where they scavenged kills and gathered plants. Some of the parasites of our ancestors followed along with them, branching as their hosts branched into new species. But hominids also picked up new parasites as they shifted to a new ecology. According to Eric Hoberg, they stumbled into the life cycle of tapeworms that beforehand had traveled between big cats and their prey.

At the same time, hominids began to spend much of their time at the few watering holes on the savannahs. There they drank from the same water that many other animals did, including rats. A blood fluke that swam from snails to rats stumbled across the skin of a hominid and tried it out. It liked what it found, and gradually a new species of fluke evolved that specialized only in hominids. Ever since then, the fluke *Schistosoma mansoni* has lived in our veins.

Hominids began moving out of Africa about a million years ago in a series of waves, hiking out across the Old World from Spain to Java. In a popular model of evolution, none of these people have any descendants left on Earth today. Instead, all living humans descend from a final wave that came out of eastern Africa a hundred thousand years ago or so and replaced every other hominid they encountered. On these travels out of the mother continent, our ancestors escaped some parasites. Sleeping sickness depends on tsetse flies to carry trypanosomes, and the flies don't live outside Africa, so sleeping sickness remained an African disease. But humans also became home to new parasites in their travels. In China, another blood fluke that had been living in rats, *Schistosoma japonicum*, moved into humans.

At least fifteen thousand years ago, some peoples headed north and east, arcing into the New World through Alaska, and there they encountered a new batch of parasites. The trypanosomes humans had left behind in Africa had existed on that continent for hundreds of millions of years. Before 100 million years ago South America was fused to Africa's western flank, and the parasites swarmed across the entire landmass. But then plate tectonics tore the two continents apart and poured an ocean between them. The trypanosomes carried away on South America began evolving on their own, into *Trypanosoma cruzi* and other species. It was long after the split between these two branches of parasites that the first primates evolved in Africa, and for tens of millions of years our ancestors struggled only with sleeping sickness. Humans migrating out of Africa escaped that scourge, but when they finally arrived in South America, the cousins of their old parasites were already there, waiting to greet them with Chagas disease.

By ten thousand years ago, humans had colonized every continent except Antarctica, but they still lived in small groups, eating animals they hunted or wild plants they gathered. Their parasites had to live according to these rules. In those early days, parasites did best if they had reliable routes into humans—tapeworms in big game, for instance, or *Plasmodium* carried by a blood-hungry mosquito, or blood flukes waiting in the water. Parasites that needed close contact might have brief flashes of glory—Ebola virus racing through a band here or there in central Africa—but the sparseness of humans didn't allow them to spread beyond that single band, so they remained rare.

That changed when humans began to domesticate wild animals and plants and eat them. The agricultural revolution sprang up independently, first in the Near East ten thousand years ago, then shortly after in China, and a couple of thousand years later in Africa and the New World. Just about every parasite boomed with the dawn of agriculture and the birth of settled towns and cities that followed. Tapeworms didn't have to wait for humans to scavenge the right carcass or hunt down the right game; they could live in livestock. After humans ate tainted pork and passed tapeworm eggs, it didn't take long for some snuffling pig to swallow them and let a new generation of parasites begin. By spreading cats and rats around most of the world, humans made *Toxoplasma* perhaps the most common parasite on Earth. Along the Andes, the houses that Incas built were ideal places for assassin bugs to live, and their llama caravans carried the insect and the parasite across much of the continent. For blood flukes, farming may have been the best thing ever to happen. With people setting up irrigation systems and rice paddies in southern Asia, huge new habitats opened up for the snail hosts of flukes, and the farmers who worked the fields were always in easy reach. Viruses and bacteria could move from person to person in the crowded, dirty conditions in the towns. And faring best of all was *Plasmodium*. The mosquitoes that carry malaria prefer to lay their eggs in open standing water, and as farmers cleared forests they brought exactly those sorts of pools into existence. The rising swarms of mosquitoes discovered new targets far more easily than their ancestors had: people toiling in fields during the day and clustering in villages at night.

For hundreds of millions of years, parasites have been shaping the evolution of our ancestors, and in the past ten thousand years they have not stopped. Malaria alone has done strange, profound things to our bodies. The hemoglobin that *Plasmodium* devours is made up of two pairs of chains, called alpha and beta, and each kind of chain is built according to instructions in our genes. We carry two genes for alpha chains—one inherited from our fathers, one from our mothers—and the same goes for the beta chains. If a mutation appears in any of those hemoglobin genes, it can damage a person's blood. One sort of mutation in the beta chain causes a hereditary disease called sickle cell anemia. In this condition, hemoglobin can't hold its shape if it's not clamped around oxygen. Without it, the defective hemoglobin collapses into needle-shaped clumps, which then turn the cell itself into a sickle shape. The sickle cells snag in small capillaries, and the blood can no longer supply as much oxygen to the body. People who inherit only one copy of this defective beta chain gene can get by on the hemoglobin made by the remaining normal copy. But people who receive two copies of the bad gene make nothing but defective hemoglobin, and they're usually dead by the time they're thirty.

A person who dies of sickle cell anemia is less likely to pass on the defective gene, and that means that the disease should be exceedingly rare. But it's not—one in four hundred American blacks has sickle cell anemia, and one in ten carries a single copy of the defective gene. The only reason the gene stays in such high circulation is that it also happens to be a defense against malaria. The needle-shaped clumps of hemoglobin don't only threaten a blood cell; they can also impale the parasite inside. And as a sickle cell collapses, it lose its ability to pump in potassium, an element *Plasmodium* depends on. You need only one copy of the gene in order to enjoy this protection. The lives saved from malaria by single copies of the gene balance off the ones lost when people get two copies of the gene and die. As a result, people whose ancestors lived in many places where malaria has been intense—throughout much of Asia, Africa, and the Mediterranean—carry the gene at high levels.

Sickle cell anemia is actually just one of several blood disorders created in the fight between humans and malaria. In Southeast Asia, for example, you can find people whose blood cells have walls that are so rigid they can't slip through capillaries. Called ovalocytosis, this disorder follows the same genetic rules as sickle cell anemia: it's mild if a person only inherits the defective gene from one parent, but severe if both parents pass it on—so severe, in fact, that a baby with two genes will almost always die before it's born. But ovalocytosis also makes red blood cells less hospitable to *Plasmodium*. Their membranes become so stiff that the parasite has a hard time pushing its way inside, and their rigidity seems to harm its ability to pump in chemicals such as phosphates and sulphates that the parasite needs to survive.

Humans have probably been fighting malaria with these sorts of changes to the blood for thousands of years, but the evidence is hard to come by. One of the few clear signs from antiquity is a condition called thalassemia, another defect of hemoglobin. People with thalassemia make the ingredients of their hemoglobin in the wrong amounts. Their genes produce too many or too few of the chains, and once the full hemoglobin molecules have been assembled from them, extra chains are left over. These end up binding together into clumps, which can wreak havoc inside a blood cell. They can grab an oxygen molecule the way normal hemoglobin can, but they can't completely enclose it. Oxygen is a dangerously charismatic element; it can carry a powerful charge that attracts other molecules in the cell. They pull the oxygen out of the defective hemoglobin clumps and carry it away. As the oxygen roams the cell, it can react with still other molecules, wrecking them in the process.

People with severe forms of thalassemia usually die before birth, but in milder forms they can survive, although often suffering from anemia. The body of a person with thalassemia may try to compensate for its defective blood cells by making more blood in the bone marrow. The marrow swells up as a result and can spread into the surrounding bone, interfering with its growth. People with thalassemia can end up with distinctively deformed skeletons—curved, stunted arm and leg bones. And archaeologists in Israel have found bones with these deformities dating back eight thousand years.

Thalassemia has lingered for so long—and has become the most common blood disorder on Earth in that time—because it helps fight malaria. If you look at a map of a malaria-prone country like New Guinea, the rates of thalassemia match up closely with the prevalence of the parasite. While a severe form of thalassemia may kill, a milder case saves. Researchers suspect that the defective hemoglobin in a red blood cell makes life worse for the parasite inside than for the host. The loose hemoglobin strands grab oxygen, which slips free and can then damage *Plasmodium*. The parasites don't seem to have any way of repairing themselves, so they can't grow properly. When *Plasmodium* finally emerges from a red blood cell, it's deformed and sluggish, and it can't invade new cells. As a result, people with thalassemia who get malaria tend to have mild cases rather than fatal ones.

These blood disorders may do more against malaria than make life hard for the parasites. They may provide a natural vaccination program for children. Children who are bitten by a *Plasmodium*-laden mosquito for the first time reach a turning point in their lives: Will their naive immune systems be able to recognize the parasite and fight it off before it kills them? Stunting the growth of parasites—whether by thalassemia, ovalocytosis, or sickle cell anemia—gives the immune system more time to get beyond *Plasmodium*'s evasions, recognize it, and mount a response. These mild cases of malaria immunize children to malaria and let them live to adulthood.

• • •

Given how much parasites have shaped the human body, it's tempting to wonder whether they've shaped human nature. Do women choose men for their parasite-proof immune systems the way a hen chooses a rooster? In 1990, a biologist named Bobbi Low at the University of Michigan reviewed the marriage systems in cultures plagued with parasites such as blood flukes, *Leishmania*, and trypanosomes. She found that the heavier a culture's parasite load, the more likely the men were to have multiple wives or concubines. You might expect that sort of result from Hamilton and Zuk's theory, since healthy men would be so highly valued in parasite-burdened places that many women would marry each one. How would women judge men for signs of parasite-proof immune systems? Men don't have roosters' combs, but they do have thick beards and broad shoulders, both of which are dependent on testosterone. The signs might not be visible either—a huge amount of communication goes on between people by odor that scientists haven't begun to decode.

If there is some connection between parasites and love, it's probably tangled up with many other evolutionary forces and slathered over with a heavy crust of cultural variations. I spoke to Marlene Zuk one afternoon about her work, which she divides between exploring the Hamilton-Zuk hypothesis and studying the songs of crickets. When I asked her what she thought of trying to apply her ideas to people, she was cautious. "It's easy to construct these adaptive scenarios and almost impossible to test them," she said. "I'm not saying people shouldn't study human behavior, that there's anything immoral about it. But I do think that there's been some shoddy work done that's gotten attention because people think, 'Isn't it cool that this thing is being applied to humans?' When people do things with humans, they get captivated with their pet theories. But I don't even understand what's going on in the structure of cricket songs."

Still, there's no crime in speculating. Could parasites have helped drive the evolution of the human mind? Primates spend huge amounts of their day—between 10 and 20 percent—grooming each other. Like other grooming animals, they have to fend off an endless assault of lice and other skin parasites. Simply picking off these parasites is soothing, because touch releases mild narcotics in the primate brain. According to Robin Dunbar of the University of Liverpool, this parasite-driven pleasure took on a new importance when the common ancestor of monkeys and apes and humans moved into habitats with lots of predators about 20 million years ago. These primates had to huddle together in order not to be killed, but they then had to compete with one another for food. As social stresses emerged, the primates began to depend on the soothing sensation that comes from grooming, not for its previous function—getting rid of parasites—but as a kind of currency to buy the alliance of other monkeys. Grooming became political, in other words, and in order to keep track of the larger and larger groups, apes evolved larger brains and had to dedicate more of their time to grooming. Hominids eventually reached the point, at about one hundred fifty members to a band, where there wasn't enough time in the day to groom one another to keep the band intact. And it was then, Dunbar claims, that language arose and took grooming's place.

Defending against parasites could have played a part in the evolution of human intelligence in another way—an even more speculative one, but one that might be more significant. Perhaps medicine played a role. When a woolly bear is attacked by a parasitic fly and switches its diet from lupine to hemlock, it does so purely by instinct. It doesn't pause on its leaf and think to itself, "I seem to have a maggot growing inside me, and it will leave me a hollow shell if I don't do something." Its tastes presumably just shift from one kind of plant to another. For most animals that engage in this protomedicine, the process is probably the same. But something different seems to be going on in primates, particularly chimpanzees, our closest relatives. Sick chimps will sometimes search for strange food. They will swallow certain kinds of leaves whole; they will strip the bark of other plants and eat the bitter pith inside. The plants have almost no nutrition in them, but they have another value. The leaves seem to be able to clear out worms from the intestines, and the bitter pith is used as medicine by people who share the forests with the chimps. When scientists have analyzed the plants in laboratories, they've discovered that they can kill many parasites.

Chimps, in other words, may be medicating themselves. As the years go by, more evidence accrues to the chimp-doctor theory, but it is slow to gain acceptance. It demands far more proof than a typical idea in biology, since scientists need to demonstrate that chimps are sick with particular parasites when they choose their plants, and they need to show how the plants fight the parasites. Proving this as you run to keep up with chimps racing along rainforest ridges makes for slow scientific progress. But Michael Huffman, the primatologist who has done most of the running, has indeed shown that after chimps eat certain plants, their parasite load drops and their health improves. He argues that the chimps are a lot more sophisticated in their medicine than instinct-driven woolly bears. When they select only the pith of the plant *Vernonia amydalina*, casting aside the bark and leaves, they are avoiding the poisonous part of the plant and taking only the part of the plant that has steroid glucosides that kill nematodes and other parasites. A hungry goat will eat too much of the plant and sometimes die.

If Huffman is right, the chimps must accrue medical lore

and carry the information through time by teaching and observing one another. Huffman once watched a male chimp eat some *Vernonia* and throw it to the ground; a baby chimp tried to pick it up, but his mother stopped him, put her foot on the pith, and carried him away. Chimps must have some remarkable cognitive sophistication if Huffman is right. They can recognize the symptoms of particular parasites and associate eating plants with their cure. They may even eat some plants preventively, which would put the association on an even more abstract plane.

You usually hear this sort of talk—abstraction, an awareness of the potential uses of things in nature—when people are discussing one of the most important steps in human evolution: the ability to make tools. Chimpanzees can strip sticks to use them to fish out termites from nests; they can smash shells between rocks; they can even fashion themselves sandals to cross expanses of low thorny bushes. As our closest primate relatives, they may embody some of the abilities of the earliest hominids 5 million years ago. Later, as our ancestors moved out of dense forests, they evolved the ability to make more sophisticated tools by flaking stones to butcher meat. The ability to connect the shape of a tool to the job it could do brought a reward of more food. This abstract thought made it possible to make better tools, and survival became even easier. Tools, in other words, may have made our brains swell.

Conceivably, that same argument could apply to medicine as well. Could the ability to recognize how plants could fight various parasites have given hominids longer lives and more children? And could that success have driven more powerful brains in order to find better cures against parasites? If that's true, perhaps a better name for us would be *Homo medicus*.

•　•　•

In 1955, Paul Russell, a scientist at Rockefeller University, wrote a book to which he gave the title—a title he thought was

The Two-Legged Host

entirely reasonable and realistic—*Man's Mastery of Malaria*. The parasite that had taken so many lives (by some counts, half of all the people who were ever born) was on the verge of succumbing to the powers of modern medicine. "For the first time it is economically feasible for nations, however underdeveloped and whatever the climate, to banish malaria completely from their borders." The end of malaria was so much a fait accompli that Russell ended his book warning that a population boom would hit the world when the parasite had been destroyed.

As I write these words, forty-four years later at the close of the twentieth century, a person dies of malaria every twelve seconds. In the time between Russell and me, scientists have unbraided the mystery of DNA; they have stared closely at the face of cells; they have climbed some of the chains, link by link, from genes to action. And yet, malaria still romps through the human race.

For that matter, so do many other parasites. Beyond the bacteria and viruses that Americans and Europeans may be familiar with, protozoans and animals are having a field day in their human hosts. There are more human intestinal worms than humans. Filarial worms, the parasites that cause elephantiasis, infect 120 million people; there are 200 million cases of schistosomiasis, the disease caused by blood flukes. Even a parasite limited in geography, such as the trypanosome that causes Chagas disease, infects close to 20 million people.

The toll taken by these parasites is overlooked for several reasons. One is that it happens mostly to the poorest people in the poorest countries. Another is that many of these parasites aren't outright fatal. Although 1.3 billion people carry hookworm, only 65,000 people actually die of it each year. But the effects of chronic infections with parasites are still devastating, leaving people listless and undernourished. Parasites like hookworm and whipworm make it hard for children to learn in school; all it takes is a dose of antiwhipworm medicine to make some slow children bright again.

Epidemiologists have tried to quantify this sort of loss with something they call the disability-adjusted life year. Simply put, this unit measures the estimated value of the years of healthy life lost to a disease. It is a grim exercise in statistics, replete with the cold-hearted calculations of labor—getting blood flukes at age twenty-five counts for much more than at age fifty-five. Depending on how bad a disease is, a year still living counted as only a fraction of a life lived parasite-free. Round-worm may slow down a child's growth, but if it's caught in time, the condition is reversed and the child begins to grow again. Left too long, though, roundworm can leave the child stunted into adulthood. When considered in this way, parasites are a staggering drain on life. Malaria robs the world's population of 35.7 million life-years every year. Parasitic worms of the gut—hookworms, roundworms, whipworms most importantly—are far less fatal than malaria but actually rob more life: 39 million life-years. Taken together, the leading parasites destroy almost 80 million life-years a year, almost twice as many as those claimed by tuberculosis.

In the United States, most people aren't aware of the damage that parasites wreak (or even know what these parasites are) because they're such a small threat to their own health today. It wasn't always the case. Most Americans don't know that in the 1800s, malaria's range swept all the way up the Great Plains into North Dakota, or that in 1901, a fifth of the population of Staten Island carried the parasite. Most don't know that people in the southern United States once had a reputation for being lazy and stupid because so many of them were being drained by hookworm. Most don't know that in the 1930s, 25 percent of the pork sold in the United States carried *Trichinella*.

The United States no longer has to worry about these parasites, but not because anyone invented a magic bullet. They've been overwhelmed by the slow, dogged work of public health, of building outhouses, of inspecting food, of treating infections to break the cycles that parasites had taken for thousands of generations before. There's still plenty of life in this simple approach. Consider the hideous case of guinea worms. Even at the middle of the twentieth century, guinea worms were fantastically successful parasites. One estimate in the 1940s had them crawling out of the legs of 48 million people every year. Today there is still no vaccine for guinea worm disease, nor is there even a medicine known to work against it. But in the early 1980s, public health workers began a campaign that may eradicate it from the face of the Earth.

Their strategy was simple. They made people in the guinea worm zone aware of the parasite's ways. They helped set up wells in some places and issued cheesecloth in other places to filter out parasite-carrying copepods from pond water. They stopped people from helping the guinea worm complete its life cycle by putting bandages on the abscesses the parasites formed. As the guinea worms were spooled out of their hosts, their hosts were kept away from water. In a matter of years the guinea worm population started to crash. In 1989, there were 892,000 reported cases (the actual cases were probably far more); in 1998, the number had dropped to 80,000. Guinea worms disappeared from Pakistan altogether in 1993. It's conceivable that within a few years, guinea worms will be completely wiped out. After smallpox, guinea worms would then become only the second disease to have been eradicated in the history of medicine.

Two other pernicious parasites also have life cycles that make them good candidates for eradication. One is *Onchocerca volvulus*, the worm that travels in black flies and causes river blindness. Seventeen million people carry the parasite, mostly in Africa. Short of wiping out all the flies or issuing insect spray to all Africans at risk, there would be no way to keep people from getting infected. Like guinea worms, *O. volvulus* has no vaccine, but it does have a partial cure. Sheep ranchers give their animals a drug called ivermectin to cure them of intestinal worms. Ivermectin seems to paralyze the worms so that they can't feed or swim, and they get flushed out of the body. Parasitologists have discovered that ivermectin actually works effectively against many other parasites, including *O. volvulus*. If a person with river blindness takes the drug, the baby worms that wander through the skin die. It's not a complete cure, since the adult worms are left snuggled happily in their nodule, where they can give birth to thousands more baby worms. But it's the babies that cause the worst symptoms of the disease—the agonizing itchiness and the scarring of the eye that leads to blindness. Researchers found that if an infected person took one pill once a year, he would be free of the babies. Since an adult worm lives ten years, he would have to take it ten times to be completely cured. The pharmaceutical colossus Merck has donated as much ivermectin as will be necessary to cure the world of river blindness, and 100 million doses have been handed out so far.

More recently, parasitologists have found that ivermectin can work as efficiently against the filiarial worms that cause elephantiasis. The filarial worms have essentially the same life cycle as *O. volvulus*, and the same susceptibility to ivermectin. The project is far more ambitious—120 million people throughout much of the tropical world are infected. If these researchers should be successful and if these three parasites are destroyed, the world should honor them for waging these campaigns. We can look forward to a time when people will have a hard time believing that there was anything on Earth that could have caused human agony in such elaborate ways. They will be the dragons and the basilisks of the twenty-second century.

Yet, in their vulnerability these three parasites are exceptions rather than the rule. Many others thrive on the poverty that most of the world lives in, and it takes more than some good intentions to stop them. Schistosomiasis is easily curable if you've got the twenty dollars to buy the drug praziquantel. If you're too poor to afford it on your own but someone gives it to you free, the chances are you'll just get sick again because you have to get your water from a pond instead of a clean well. And often the supposed cures for poverty make the lives of parasites easier. When giant dams are built and submerge vast regions of dry land, they create new homes for the snails that carry blood flukes, and new epidemics of schistosomiasis reliably follow.

The most important reason that parasites do so well today is that they evolve. Parasites are not life's dead ends, as was once thought; they are continually adapting to their circumstances. Not only has malaria been forcing us to evolve; it has been evolving to adapt to us. And after adapting to natural human defenses for many thousands of years, *Plasmodium* now simply has to go up against drugs rather than some new T cell receptor.

Before the 1950s, the malaria a person contracted anywhere in the world could be treated with a few doses of the benign drug chloroquine. Chloroquine cures malaria by turning *Plasmodium*'s food into poison. As *Plasmodium* feeds on the hemoglobin in red blood cells, the parasite chops off the arms of the molecule, leaving behind the iron-rich core. This core is dangerous to the parasite, because it can lodge in *Plasmodium*'s membrane and disrupt the flow of molecules in and out. The parasite neutralizes the poison in two ways. It strings some of the molecules into harmless hemozoin; the rest it processes with enzymes until it can no longer react with the membrane.

Chloroquine works its way into *Plasmodium* and bonds with the hemoglobin core before the parasite can neutralize it. In its new form, the compound won't fit on the end of a hemozoin chain, and the parasite's enzymes can no longer react with it. Instead it builds up in *Plasmodium*'s membrane and makes it leaky. The parasite can no longer pump in the atoms like potassium that it needs, or pump out the ones that it has to get rid of, and it eventually dies.

Now huge parts of the globe harbor malaria that's chloroquine-proof. In the late 1950s, two chloroquine-resistant parasites were born—one in South America, the other in Southeast

Asia. Researchers aren't exactly sure what makes them so stubborn, but they suspect that they have a mutant protein that snags chloroquine before it penetrates too deeply into the parasite. These mutants have probably cropped up regularly for thousands of years, but the odd proteins they produced served no good purpose. They probably even slowed down the parasite's feast of blood, so they were squelched by natural selection.

But starting in the 1950s, any parasite that could block chloroquine had plenty of space—human bodies—for colonizing. Year by year, the children of those two *Plasmodium* mutants spread from their homelands. The South American mutant spread to cover every malarial region of the entire continent. The Southeast Asian mutant, meanwhile, was even more cosmopolitan: by the 1960s it had overrun Indonesia and New Guinea to the east, while to the west it spread in the 1970s through India and the Middle East. In 1978, the first record of this Southeast Asian form was recorded in East Africa, and in the 1980s it had made its way to most of sub-Saharan Africa. Now it's much harder to stop the spread of malaria because other antimalarial drugs are more expensive, and resistant strains of *Plasmodium* are rising up against them as well.

The resurgence of parasites like *Plasmodium* has made parasitologists yearn for a vaccine. But even though vaccines work well against some viruses and bacteria, there's no commercially available vaccine against a eukaryote. None. The problem is that eukaryote parasites are complex, evasive creatures. They go through different stages within their host, one stage looking nothing like the next. Protozoans and animals are accomplished at fooling our immune systems—just consider the way trypanosomes can peel off their molecular fur and grow one with a completely different pattern of chemical stripes, the way blood flukes snatch our own molecules for a mask while producing other chemicals that turn us against ourselves.

The first attempts to make parasite vaccines were crude affairs. Scientists would destroy live parasites with radiation and then inject their remains into lab animals. They provided only a little protection. In the last twenty years, scientists have learned how to tailor their vaccines much more carefully. They've turned their attention from entire parasites to single molecules the parasites carry on their coats. Their hope has been to find a handful of molecules that the immune system can use to prime itself for fighting these invaders. But still the failures have kept coming. The World Health Organization organized an aggressive campaign to create a schistosomiasis vaccine in the 1980s. They backed not one molecule but six, each tested by a squadron of immunologists. None of them offered any significant protection, so the grand scheme has been scrapped as the vaccine developers look for new molecules.

Yet, parasites do not by definition defy vaccines. It's still possible that there is a molecule they simply can't live without, that the immune system can identify regularly enough to use as a guide for their attacks. In 1998, human trials began for a vaccine for malaria created by scientists with the United States Navy. Their vaccine is even more sophisticated than current ones. They want to get the human immune system to attack *Plasmodium* at its early stage in the liver cell. The liver cells display bits of *Plasmodium*'s proteins in the receptors for major histocompatibility complex (MHC) on its surface. Normally our bodies can't fight malaria at this stage, because by the time killer T cells have recognized the fragments and multiplied into a parasite-killing army, the *Plasmodium* has already escaped the liver and slipped into the bloodstream.

But if the killer T cell were already primed to recognize those fragments, they would be able to start destroying the infected liver cells immediately. To create an army of these T cells, the navy scientists want to give people a false case of malaria. They have fashioned a sequence of DNA that they are injecting into the muscles of volunteers. The DNA makes its way into the muscle cells, where it starts making the same protein that is made by *Plasmodium* and displayed by liver cells. The muscle cells should, in theory, carry this vaccine protein to their own surface, and killer T cells that come across it will be able to fight off an actual infection when it comes.

It's a long way, though, from human trials to an actual vaccine campaign—particularly against diseases such as malaria and schistosomiasis that affect hundreds of millions of people in the poorest parts of the world. "What's the best you could expect from a vaccine?" asks Armand Kuris, who has spent a large part of his career looking for ways to control schistosomiasis. "A molecular biologist will say, 'It's expensive, it will require revaccination every five to seven years, it will require perfect cold delivery.' That means refrigeration from its manufacture to the point when you're taking out a vial and sticking a syringe into it. Did you ever get a vaccination for smallpox? I received a vaccination on the border of Costa Rica where the nurse had the vaccine in a shot glass and tattooed me with a sewing needle. Now *that's* a vaccine." He points out that praziquantel, the cure for schistosomiasis, costs twenty dollars. "In Kenya in the villages where I work, the best-off families may be able to get the drug for a favored child. If that's economically impossible, then if I gave you a vaccine, what the hell could you do with it? I'm not saying don't do any research in it. The navy may have to go to a place with malaria—Peace Corps workers, diplomats . . . but in terms of the 200 million people who suffer from schistosomiasis, the vaccine has no chance of working. And yet my calculation is that three-quarters of the money spent on schisto in the past twenty years has been spent on vaccines."

Even if researchers could produce a vaccine that met Kuris's shot-glass standard, the parasites might well find a way around it. The World Health Organization has decided that even if a schistosome vaccine provided only 40 percent protection, it would be worth backing. That doesn't mean that 40 percent of the 200 million people with schistosomiasis would be rid of their parasites. That means that each person would lose 40 percent of the worms inside his veins. It sounds like a worthy goal, but it ignores the sophistication of schistosomes. These flukes can sense how many of their fellow flukes are in their host, and as that number gets higher, each female produces fewer and fewer eggs. It's probably a mechanism the blood flukes have evolved to take care of their hosts. If every female were to crank out as many eggs as she possibly could, they'd cause so much scarring to the host's liver that the host might die. A vaccine that killed 40 percent of the worms in a person might create the opposite situation: the surviving schistosomes would sense that they had less competition and ratchet up the egg production, making the disease worse.

Vaccines also run the risk of tearing down our hard-earned ability to immunize ourselves. Say that the navy vaccine against the liver stage of malaria works, and that it is decided to inject it into millions of children around the world. Now say that the vaccine works brilliantly for a few years. Now say that countries let the program lapse because of civil war or because speculators sell off the national currencies. Or, if you like, say that a mutant strain of malaria sweeps through, different enough to keep T cells trained on the vaccine from recognizing it. Now the people would have no protection in their livers, and wouldn't have had the opportunity to build up their own resistance to the blood stage of the parasite. The vaccine could then conceivably cause more harm than good.

For some parasites, it may actually make more sense to find a better coexistence than to try for eradication. In schistosomiasis, for example, the adult blood flukes themselves don't cause much harm. They're so well cloaked from the immune system that they don't trigger a damaging attack, and they don't drink much blood. It's their eggs that are trouble, as the immune system forms giant balls of scar tissue around them in the liver. Among the many signals that immune cells trade, one has the ability to stop them from making these granulomas. Scientists have found that if they give an extra dose of this signal to mice with schistosomiasis, they don't destroy their own livers. Conceivably, this kind of medicine could save us—not from parasites but from ourselves. Another strategy could be to keep the blood flukes from mating. Scientists have discovered that males attract females with a chemical signal. If people were vaccinated so that their immune system could destroy that signal, blood fluke love would be foiled, and no eggs would be made.

Coexistence with parasites might also be possible if we could tame them. The severity of a disease caused by a parasite has a lot to do with its evolutionary options. If a virus's best chance for survival requires it to kill its hosts quickly, it will probably evolve into a lethal strain. But the opposite is also true: if the virus has to pay a heavy price for being virulent, more benign strains will win out. For well over ten thousand years, we've actually been managing a lot of evolution as we've bred plants and animals for the qualities we desire—docile cows, for example, and sweet apples. One of the architects of the theory of virulence, Paul Ewald of Amherst College, has proposed doing the same thing with parasites in order to fight diseases. It's actually not hard to domesticate a parasite. In many parts of the tropics, for example, public health campaigns are supplying people with screens and bed nets to keep malaria-carrying mosquitoes from biting them as they sleep. The campaigns will save lives not only by preventing mosquito bites, Ewald suspects, but by forcing the *Plasmodium* inside the mosquitoes to evolve into a gentler form. As it becomes less likely that a parasite can get from one host to the next, it becomes unwise, evolutionarily speaking, to kill a host.

Eradicating parasites may even create new diseases. Colitis and Crohn's disease affect 1 million Americans today. In both cases, a person's own immune system violently attacks the lining of the intestines. The inflammation it triggers ruins a person's digestion, and sometimes a surgeon may have to cut out a length of the damaged bowels. Both diseases can torment a person for a lifetime, and so far there's no cure for either. Yet, as common as they are today, you can't find any record of colitis or Crohn's disease before the 1930s. The first cases in the United States turned up in well-to-do Jewish families in New York City, which made doctors think they were hereditary diseases. But then whites who weren't Jewish started getting them. Still doctors thought the diseases were hereditary because hardly any blacks fell ill. But in the 1970s, blacks started getting the diseases as well. Looking outside the United States, you can see another peculiar pattern. In the poorer countries in the world, the diseases are practically unheard of. Yet, in Japan and Korea, two countries that have quickly gone from poverty to wealth, there are now epidemics of colitis and Crohn's disease.

Some scientists think that the spread of these diseases was caused by the eradication of intestinal worms. The idea certainly fits their history. In the United States, they appeared first in affluent people in cities—the people, in other words, who would have been the first to be cleared of tapeworms, and other worms living in their bowels. Later, when blacks began to emerge from poverty and moved to cities as well, they also fell ill. Intestinal parasites are still common in most of the world, but in countries where they've been recently eradicated, colitis and Crohn's disease have followed fast. Even farm animals are starting to get bowel diseases as they've been getting treated with antiworm medicines like Ivermectin.

Humans may have been protected from diseases like these by the interplay between their immune systems and intestinal parasites. Parasitologists have found that intestinal worms can nudge the immune system from a poison-spouting, cell-engulfing frenzy to a gentler sort of attack. In this mellower mood, the immune system can still keep bacteria and viruses in check, but the parasitic worms can live unmolested. This arrangement benefits the host as well. When parasitic worms are abundant, it would be dangerous to attack them over and over again. But then, in an evolutionary blink, a few hundred million people lost their parasites. Without their soothing influence, some people now swing too far the other way, their immune systems unable to stop attacking their own bodies.

In 1997, scientists at the University of Iowa put this idea into startling practice. They picked out seven people with ulcerative colitis and Crohn's disease, who had gotten no relief from any conventional treatment. They fed them eggs from an intestinal worm that normally lives in an animal, one that wouldn't cause any disease of its own in a human gut (the scientists are still keeping the species a secret until they've finished their research). Within a couple of weeks the eggs had hatched, the larvae had grown, and six out of the seven people went into complete remission.

Parasite-free living may also be responsible for the rise of other immune disorders, such as allergies. Twenty percent of the population of the industrialized world suffer from allergies, but elsewhere they're hard to find. Since it's dangerous to generalize from country to country, an immunologist named Neil Lynch has done fine-grained studies of this pattern in Venezuela. He looked at people in upper-class homes with running water and toilets, and compared them with poor Venezuelans in slums. While 43 percent of the upper-class people had allergies, only 10 percent had light infections from intestinal worms. Among the poor, there were half the allergies as in the upper classes but twice the worms. And when Lynch studied Venezuelan Indians who live in the rain forests, the pattern was even starker: 88 percent were infected with parasites, and they had no allergies at all. Without parasitic worms exerting their influence, our immune systems may be prone to overreacting to harmless bits of cat dander and mold.

To fight these diseases, we may need to acknowledge our long marriage to parasites. That's not to say that people with colitis should be eating *Trichinella* eggs unless they'd enjoy a long, agonizing death as the parasite worked its way into their muscles. But the chemicals that the parasites use to manipulate our immune systems may offer protection from modern life. Perhaps some day, along with polio vaccines children will get parasite proteins, so that their immune systems will be trained not to fly out of control. It would be a supreme final twist to the story of parasites in humans. They may not always be the disease. In some cases they may be the cure.

8

How to Live in a Parasitic World

Whenever the earth changed its form of existence, the existing creations were also destroyed. The same thing occurs to the worms; when the host animal dies they are also destroyed.

— Johannes Bremsner, German parasitologist (1819)

On my visit to Santa Barbara, after Kevin Lafferty had showed me how parasites hold sway over a salt marsh, I spent a morning with one of Armand Kuris's graduate students, a young man named Mark Torchin. He led me through one of the marine biology labs to a blue door in the corner. A sign marked QUARANTINE plastered the door. When Torchin opened the door and we walked into the dark, I could hear what sounded like a flowing creek. Torchin found the switch to the cold fluorescent lights, which shone down on a high table running the length of the room.

How to Live in a Parasitic World

On the left side were aquarium tanks full of water, with crabs skittering around inside on broken pieces of white mesh. On the right were tubs with cups stacked in them, each holding a single crab in a scoop of water. The sound of a creek came from the system of pipes that pumped sea water in from the lagoon just outside, flowing into the tanks and dribbling onto the table before heading down a drain, to flow back to the Pacific.

The crabs were *Carcinus maenas*, the European green crab. Some were the size of teacups, some only of shot glasses. If you walk along the coast of northern California and the Pacific Northwest, you may find green crabs, and that's a fact that has certain people terrified. Before 1991, there were no green crabs on the California coast. Its original range was along the beaches of Europe. There it was a voracious creature; in Great Britain, biologists have watched single crabs eat forty cockles, each half an inch long, in a single day. For thousands—perhaps millions—of years, the rest of the world was spared from the green crab's hunger, but that changed when humans invented ships. The green crab sheds thousands of nearly invisible larvae into the water, which can be easily sucked into the holds of ships when they take on ballast water. Perhaps two hundred years ago, some ship traveling to the American colonies carried green crabs to the New World. They quickly began to spread along the coast of the eastern United States, devouring shellfish in northern New England and Canada. The softshell clam, once the basis of a whole fishing industry in New England, disappeared altogether.

The crabs traveled to South Africa and Australia as well, but for centuries the west coast of the United States was spared. Despite all the ships traveling there from Europe and the eastern United States, it wasn't until 1991 that a fisherman near San Francisco pulled up a green crab in his nets for the first time. As soon as reports spread around marine biology circles, scientists became gloomy. Almost every species of shellfish around San Francisco was suitable prey, and if the green crab should spread along the coast in the ships that traveled down to Los Angeles or up to the northwest, it could spread to new habitats, feasting on oysters, Dungeness crabs, and other valuable creatures. The burrows it dug might destabilize dikes, levees, and channels, causing even more damage. "It's a disaster," says Armand Kuris. "It's all the things you want in a worst-case scenario."

The green crabs in the quarantined lab in Santa Barbara skittered in their tanks. Some had ghostly white claws growing in the place where they had lost a previous one. And some, as I could see when Torchin pulled them out of the water and turned them upside down, their legs and claws windmilling around helplessly, carried a sac on their abdomen the color of butterscotch. They looked like normal crabs, but they had been transformed into something else. They were filled with *Sacculina carcini*, that degenerate parasitic barnacle of Ray Lankester's nightmares. Torchin, Lafferty, and Kuris were trying to use *Sacculina* to save the Pacific coast from the green crab.

In the late 1800s, scientists sometimes referred to parasitology as medical zoology. They were referring to the way they had to understand parasites as real organisms, with natural histories of their own, before they could try to fight the diseases the parasites caused. Now, a century later, the term has taken on a new life. Now the patient isn't a person but the natural world. Alien species are spreading uncontrollably across continents and seas; native plants and animals are falling prey to new diseases; habitats are disappearing as forests turn to stumps and coastlines to condominiums. As ecosystems have faltered, scientists have come to recognize that parasites are important to their health. A healthy ecosystem is riddled with parasites, and in some cases, an ecosystem may even depend on parasites for its health. As humans alter the world, tipping the biosphere out of kilter, it may be possible to enlist parasites to help us undo some of our mistakes and perhaps keep us from making new ones.

Scientists first conceived of using parasites against pests in the 1880s. The original idea was simple. A parasite is a cheap, never-ending pest-killer. It can seek out its host and invade it, fighting off the host's immune system and, in many cases, leaving the host dead. Farmers who use pesticides have to spray their plants at least once a year, but parasites keep regenerating and tracking down new hosts. Simply sow the parasite, the argument went, and your troubles are over. In the early part of this century, farmers were having exactly the sort of success that had been promised. Scales and beetles and other pests were destroyed by wasps and flies and other sorts of parasites. The parasites couldn't eradicate the pests completely, but they no longer threatened to wipe out whole fields.

In the 1930s, the agrochemical industry was born. DDT arrived on the market, a powerful pesticide that came with the luster of modern science—a synthetic creation that humans could use to master nature. As a result, biological control withered away. A few biologists in California and Australia kept studying parasites in the hope of bringing back biological control. And over the next forty years, pesticides began to falter. Insects evolved resistance to DDT. The chemical worked its way into the food chain, causing birds to lay eggs with thin eggshells. An environmental movement opposed to pesticides started up, and the aging masters of biological control saw a chance for a comeback.

"I was a graduate student at Berkeley at the time," says Armand Kuris. "It was so interesting. These were *old* guys, twenty years, thirty years my senior. They were old agricultural guys with string ties and stuff like that. And there they were in the sixties with all the hippies, and they found themselves in the same bed together. In the beginning it was weird, but then they realized they *were* on the same side. It was one of the sidebars to the history of the sixties."

In its second incarnation, biological control with parasites had a much more solid scientific foundation. Insects can evolve resistance to DDT, but parasites can evolve as well. They can come up with new molecular formulas for attacking their hosts, canceling out any resistance the pests may evolve. A parasite could rein in a pest, some scientists argued, by bringing back at least some balance to nature. Most pests are alien species like the green crab, brought to a new land. One reason they are so harmful is that they have escaped their parasites and can breed unchecked, while native species have to struggle against their own parasites. Introducing a parasite from the invader's homeland, the argument for biological control goes, is really just a way to reestablish some natural restraints.

• • •

The new biological control has in fact produced some spectacular triumphs over dangerous hosts. It may, for example, have saved much of Africa from starvation. What rice is to China, what potatoes once were to Ireland, cassava is to Africa. The plant grows three feet high, with broad green leaves that are as nutritious as spinach and far tastier. The roots of spinach don't count for much, but cassava roots are thick slabs of starch. Cassava is rugged enough to grow where other roots would rot away, so for some villages in the wetter parts of Africa it's the only thing poised between them and famine. From Senegal, on the Ivory Coast, to Mozambique, on the Indian Ocean, 200 million people depend on it. And in 1973 the cassava began to die.

On the little plots around Kinshasa, the capital of Zaire, leaves began to curl and shrivel, and without photosynthesis the roots became stunted. Within a few years there was so little cassava around the city that a family's supply for a week cost more than a month's wages. In the meantime, cassava began to die around other port cities along the Atlantic coast of Africa: Brazzaville, Cabinda, Lagos, Dakar.

When people uncurled the withered leaves, they found a white speckling, which resolved itself under a magnifying glass into thousands of pale flat insects. No one had ever seen the insects before in Africa; in fact, no one had ever seen this particular species before anywhere in the world. Known as cassava mealybugs, they are one of the many plant-eating parasites, tuned to the narrow frequency of their host-plant species. The insect stabs the cassava leaf with its proboscis, which anchors it in place. It sucks out the sap, at the same time injecting a poison that somehow stops the roots from growing, which probably lets the mealybug take up more food through the plant's leaves. Cassava mealybugs are all female, and a single female can lay eight hundred eggs in its microscopic lifetime. By the end of a growing season a single shoot may sag with twenty thousand insects.

The curling of the leaves is also caused by the mealybug's poison. It may be that the shriveling helps the insect spread from plant to plant. A healthy cassava field puts up a thick blanket of leaves to the wind, deflecting breezes up and over the plants. But when cassava becomes host to mealybugs, the blanket becomes tattered, letting the wind work its way among the shoots, carrying with it young larvae to colonize new plants. While this is only a theory, there's no doubt that once a single cassava plant in a field falls to the mealybug, the rest are doomed. To make matters worse, cassava is a portable plant; a farmer can take a shoot and start a new field with it somewhere else. If even a single mealybug is hidden in the leaves, the new field, and the older fields around it, become infested.

The leaping of the mealybugs from port to port was probably brought about this way. Someone may have even taken a mealybug on a plane, because in 1985 it turned up several thousand miles away in Tanzania, where it began to spread from field to field. Wherever it went, it didn't simply rob farmers of a single year's crops. Since they needed cuttings to replant their fields, and none of their cuttings was free of the mealybugs, the farmers lost the crops for years to come.

In 1979, a Swiss scientist arrived in Ibadan, a Nigerian university town deep in cassava mealybug country. He was Hans Herren, an entomologist who had grown up working on his family's farm outside Montreux. "As I was growing up, we were going from almost completely organic farming to a full pesticide thing," Herren told me twenty years later when I visited him in Nairobi. His hair had gone gray, but he was still a live wire, able to tell a story rapid fire for an hour straight. "I can remember in ten years going from using almost no chemicals to using herbicides and pesticides. I was the one driving the tractor off hours from school, treating our potatoes, our tobacco, our wheat, and everything else with all these chemicals. I remember these guys coming around the farm selling chemicals to my father. I saw how we did it before, and then we went into this treadmill of more and more and more."

Herren went to college hoping to find a way to jump off the treadmill without landing too painfully. He studied biological control, first in Switzerland, then at the home of its renaissance at the University of California at Berkeley. The International Institute of Tropical Agriculture offered him a job, or, more precisely, a challenge: Could he find a parasite for the cassava mealybug? He didn't think twice before taking the job. "Going to Nigeria was a chance to practice on a very large scale what I had learned in Berkeley and Zurich."

When Herren arrived at Ibadan, he discovered that most of the scientists there were sure he would fail. They were breeders, creating new cassava hybrids designed for fast growth and resistance to disease. They were sure they could handle the mealybug disaster. "They said, 'Mealybug? No problem: breeding, that's the solution.'" And when they met Herren, their thoughts ran in a different direction: "'This guy from Berkeley—what does he know? This ecological freak.'" Herren himself had nothing against breeding, but for the crisis at hand there simply wasn't enough time. The mealybug was catapulting from one city to another and then racing through the surrounding farm land "like a dust cloud," says Herren. Breeding a resistant hybrid can take a decade, and in ten years there might not have been any cassava left to save.

How to Live in a Parasitic World

In order to find a parasite for the cassava mealybug, Herren had to find where the mealybugs had come from. They had appeared out of nowhere around Kinshasa. They were not related to any known mealybug in Africa, but to a species that lived on cotton across the Atlantic, in the Yucatan. "Then I started to think, 'Well, it's from Central America—that's interesting, because cassava is also from the Americas originally. The Portuguese brought it to Africa back in the slave trade. The voyage was a very long one, down in the ship, and the salty water killed whatever was on it, so they never brought any insects across. So the plants were really thriving for several hundred years until somebody brought in mealybugs." No one had ever seen the cassava mealybug in the New World, Herren reasoned, because there was some parasite there keeping it at bay. "If it were not under control we would already know about it."

Herren paged through entomological and agricultural journals, reading up on the insects that ate domesticated cassava. "Something didn't make sense. The scientists in the Americas had been working on cassava for the last fifty years, breeding, all kinds of things, and nobody had seen that mealybug. Now wild cassava, a lot of them are used as ornamentals. They are the most beautiful plants. So I thought, maybe somebody carried a nice-looking plant. If nobody has found this mealybug in the cassava plants in so many years, why should it be there? I was going to have to look not only at cassava but at its wild relatives."

Looking throughout Latin America for an insect no one had seen before would take even longer than trying to breed cassava out of its woes. But throughout the range of wild cassava, Herren recognized a few hot spots of cassava genetic diversity. They might also be where the most diverse of cassava-eating insects are. And one of those insects might turn out to be the one eating up Africa.

Herren set off for the Americas in March 1980. He started by visiting several museum collections of plants, looking at dried specimens of cassava. It was possible, he thought, that someone had already found what he was looking for. "But I could find nothing, so I said, let's go look for the real thing. I went over to California and bought myself a big van. I established a lab in the back, a bed, everything, and I started driving through Central America, all the way to Panama, looking for wild cassava and cultivated ones."

As Herren wandered down through Central America, a network of entomologists there was also on the lookout for the insects. Many new mealybugs turned up in the search, but none of them turned out to be the species blooming in Africa. "I decided, okay, let's go away from Central America. Let's go to South America. I parked my van in the Panama airport and flew down to Colombia to visit a friend of mine. We set off to Venezuela and looked at one of the centers of diversity, the northern part of Venezuela. We drove for weeks. We found a lot of cassava mealybugs, but never the right one. So I gave him pictures, good photographs of what I was looking for, what the plant looks like when the mealybug is on it, and I went back to Africa."

His friend Tony Bilotti went to Paraguay not long after Herren went back to Ibadan. He was visiting some fellow Americans serving in the Peace Corps, and he knew that he was now in a cassava hot spot in Latin America, the only one that Herren hadn't had time to visit. Driving one day past a field of cassava, he noticed a few plants that looked a little funny. He stopped and plucked the leaves. Inside them he saw Herren's mealybug.

When Herren got word, he had Bilotti send the insects to the British Museum, where entomologists could identify them precisely. Although the insects were dead, the entomologists recognized them as the species in Africa. And as they dissected them they discovered inside their bodies the true end of Herren's search: parasitic wasps. Now Herren had the parasite that kept the cassava mealybug a minor pest in one corner of

Paraguay, and the parasite he needed for Africa. He had entomologists in Paraguay send live mealybugs to England, where
they could be raised under quarantine and the parasites could
be captured as they emerged from their hosts. He sent mealybugs and cassava plants from Africa to the same quarantine,
where scientists were able to get the wasp to lay its eggs in
them. Even more important, the experiments showed that the
wasps could lay eggs *only* in the cassava mealybugs. They hadn't
tuned themselves to the immune systems of other mealybugs,
which could choke the wasp eggs in suffocating capsules. The
wasps, Herren decided, would be safe to bring to Africa. Three
months later, Herren got his first shipment of the wasps.

He was ready for them. He and his students at Ibadan had
been building greenhouses where they could grow cassava infected with mealybugs and capture the wasps that thrived on
them, and they figured out how to mate the wasps. After they
had collected a few hundred of the egg-laying females, they
made their first release in the fields around the Ibadan campus
in November 1981. "Within three months, the mealybug population crashed. Then we knew we had something good going.
It was barely a year and a half that we had gone from not knowing anything about this to having something in the field that
worked."

Biological control, even in its renaissance, remained a mod-
est enterprise. Entomologists would raise wasps in their labs
and load them into small containers that they'd take with them
when they drove to orchards or corn fields. But a great dream
took possession of Herren: to spread the wasp across Africa.
"What I didn't like in biological control was the way it was done
as a shoestring operation, in a cheap way, using a secondhand
beaker, raising wasps in some small cages—not done in the best
possible way. That's why biological control lost to chemicals."

He knew that the dream would be expensive: $30 million, in
fact. "That was when I was called a megalomaniac. I said,
'Look, when you guys over in America have a fruit fly outbreak

in California, which is only the size of a pin compared to this whole thing over here, you spend $150 million in one year. We're talking about 200 million people who are at risk, not a few businesses that make oranges. We are dealing with one and a half times the area of the United States. We're not going to do this in cages and on donkeyback and bicycles. We're going to do this with technology, machinery, electronics, aircraft.'"

Maybe it was the word *aircraft* that made people suspicious. Herren claimed that he would be able to spread his wasp across Africa by sowing it, crop-duster fashion, from a plane. The wasps were put to sleep with carbon dioxide and then lodged in cylinders of foam rubber, two hundred fifty in each, which were loaded into a magazine that had been custom-built for Herren at an Austrian camera factory. As the pilot passed over a field Herren intended for him to drop the wasps precisely. "It was like in fighter aircrafts. You know when to drop the bomb by looking at the crosshairs. We tried this over a swimming pool in Ibadan. We'd fly over and drop the wasps. At one hundred eighty miles per hour, we were able to get them in there."

In the meantime, the wasps Herren had set free in the fields around Ibadan had been thriving. Two years after their release, he decided to see how far they had spread. "We went on foot. We thought, 'Oh, no big deal, we'll just walk.' And we walked the whole day, and we kept finding them. We thought, there's something wrong here. Nobody had ever seen this sort of wasp spread more than a few kilometers. And the next day we came back and we took the car and we drove. We drove one hundred and fifty kilometers before we finally found no more wasps."

By 1985, thanks to these early successes, Herren had collected $3 million of start-up money, and his pilots were strafing the countryside with wasps. The parasites tumbled out of his plane and landed on fields in Nigeria, in Kenya, in Mozambique, in countries from the Atlantic Ocean to the Indian Ocean. His team was raising 150,000 wasps every month, and although many of the wasps died during the long journeys from

How to Live in a Parasitic World

Ibadan to the release sites, he really needed only a single viable female wasp to survive the flight and the fall and to start looking for hosts. Even among parasitic wasps, the host-hunting skill of the Paraguayan species was extraordinary. "The wasp has developed an ability to search which is fantastic," Herren says, with a pride that is almost paternal. "If you have one plant with mealybugs on it in a field that's a hundred meters by a hundred meters, the wasp will find it. We tested this. We had fields that were clean. We put mealybugs on one plant, and we released the wasps from a corner of the field. Within a day they were on the plant. Then we tried something else. We put the mealybugs on the plant and then took them off. Then we released the wasps and they ended up on the same plant. There's something that the plant releases that attracts the wasps, a cry for help."

Herren trained twelve hundred people from the countries where the wasps had been introduced to recognize it. A few months after the drops, they began to comb through the fields to see how fast the wasp was spreading and how the mealybugs were faring. "Everywhere the problem was gone twelve months after the release. We could hardly believe it ourselves, that it worked so fast."

The last flight of the wasp duster was in 1991, but for the next few years entomologists still went on tracking its effects. In about 95 percent of the fields where the wasp had been released, the mealybug had virtually disappeared. As they lost their hosts the wasps had diminished to only a few survivors as well. In the remaining 5 percent of the farmland, the mealybugs still thrived, but Herren was able to show why: the farmers didn't take good care of their fields. Their plants were scrawny, and the mealybugs that fed on them tended to be scrawny as well. The species of wasp that Herren used is a careful judge of the size of its host, able to use its antennae like a ruler to figure out how big a mealybug is. Only then do they decide which sex to make their offspring. (When a female wasp mates, it stores the male's sperm in a gland, which it can use later to fertilize its eggs. Thanks to wasp genetics, an unfertilized egg will grow up to be male, while a fertilized one will grow up to be female.)

The wasps choose to lay only males in small mealybugs. Their logic lies in the cheapness of males. The chances of an egg successfully maturing to an adult are worse in a small mealybug because there's less food for the parasite to eat. Because the wasps put males in small hosts, only a few of them may survive to adulthood. But that doesn't matter because it takes only a few males to inseminate a lot of females.

Thanks to the wasp's strategy, a field of badly farmed cassava will be filled with male wasps. Since males don't lay eggs, they pose no threat to the mealybugs, which have a chance to quickly rebuild their population. "We've told the farmers, 'Look, biocontrol can only work when everything else is in good shape. If you don't weed your field, nothing will work.'"

Herren told me the story of the cassava mealybug one sparkling day in Nairobi. He had moved there in 1991 to become the director general of the International Center for Insect Physiology and Ecology, a massive complex on the outskirts of the capital with sculptures of dung beetles out front. The job is one of his many rewards for having saved the staple crop of 200 million people. The center is filled with entomologists trying to find ways to use insects to make human life better by producing honey and silk and by destroying pests. A stem borer has been chewing its way through the corn of East Africa, but Herren's scientists have found a wasp from India that parasitizes it. When I visited, they had already set it loose in Kenya to see whether it would survive in the wild. It did, and now they had no idea how far it had spread. And that sort of ignorance was fine with them.

• • •

Lafferty and Kuris wanted to do for the green crab what Herren had done for the cassava mealybug. They knew that in Europe many green crabs were plagued by parasites such as

Sacculina, but the crabs they dissected from San Francisco Bay were parasite-free. That might be one of the reasons why it could outcompete other crabs in its new home. So Lafferty and Kuris began to contemplate bringing *Sacculina* to California as well. *Sacculina*-infected green crabs could be dropped into the Pacific waters. They would act like miniature parasite crop dusters by spraying *Sacculina* larvae into the water. The larvae would seek out uninfected crabs, burrow into them, and spread their tendrils out. Bringing *Sacculina* to California wouldn't have the same effect as the parasitic wasps had on cassava mealybugs, because the biology of the two parasites is very different. The wasp kills its hosts by devouring their innards and then chewing its way out of their bodies. *Sacculina* doesn't kill its green crab hosts, but it does castrate them and then make them compete for food with uninfected crabs. Lafferty built mathematical models that suggested that if *Sacculina* came to the Pacific, it would make the crabs decline, but more slowly than the cassava mealybugs. It would be the missing crab eggs that would bring down their numbers, rather than dead crabs. So when *Sacculina* and the green crab finally reached an equilibrium with each other, the crabs would be reduced but not gone.

But to Lafferty and Kuris, it didn't seem as if there were any other choices. "All other alternatives are way worse ecologically," says Kuris. "Antibarnacle paint on boats is polluting our estuaries in a major way. Up in Oregon there's someone in a crop duster spraying mud flats against ghost shrimp, to protect the god-damned *introduced* oyster production. It's killing Dungeness crabs."

For a few years, Lafferty and Kuris couldn't drum up any funds to study *Sacculina*, but by 1998 the green crab had reached the shores of Washington State. It was poised to move into Puget Sound, with its huge Dungeness crab fishery. At last Kuris and Lafferty got the money they needed. They contacted the world's expert on *Sacculina* and related parasitic barnacles, a scientist in Denmark named Jens Høeg. Høeg sent them coolers filled with infected green crabs packed in ice.

Mark Torchin, Kuris's graduate student, set up the crabs in a quarantined room. He couldn't simply seal off the room completely, though, because the crabs and the parasites needed circulating sea water to survive. Torchin built pipes that pumped sea water in from the ocean; the water poured into a group of tanks, and the overflow, which might carry the invisible parasite larvae, traveled through a series of filters and tubs of gravel before pouring into an outgoing pipe headed for a nearby lagoon.

For months, Torchin slowly got acquainted with *Sacculina* and its bizarre life cycle. He figured out how to recognize when a crab was getting ready to release a new batch of parasite larvae from the sac on its abdomen (the sac would turn from butterscotch-colored to a dull caramel). He would put the crabs in little plastic cups to collect the larvae, and then he'd siphon off some of the *Sacculina*-laden water. He would pour it into another cup with a healthy green crab and wait for the female *Sacculina* to get into its new host.

Each day he would grab a crab by the claw and pinch it with his fingers. To escape, the crab would sever its own limb from the inside and drop back into the water. Torchin would take the limb to his microscope and look for larvae grabbing onto the hairs of the crab's claw and digging into the soft pits that anchored them. When a female *Sacculina* succeeded in infecting a crab, he'd let it develop into a knob on the crab's abdomen, and then he'd try to get males into it.

After a few months, Torchin was able to shepherd *Sacculina* from larva to adult. Then, at the beginning of 1999, he applied what he had learned to native California crabs. He chose the common shore crab, *Hemigrapsus oregonensis*, and exposed it to *Sacculina*. This was probably the first time in the history of these two species that they had ever met—a crab from California and a parasitic barnacle from Europe. Torchin waited to see what would happen.

A female *Sacculina*, he discovered, had no trouble getting inside the shore crab. It could even send its tendrils out through its new host's body. But then something went wrong. In a European green crab, the parasite can carefully wind its tendrils around the nerves, not only causing no damage to them but passing mind-altering signals to their host. In the shore crab, though, *Sacculina*'s tendrils just seemed to destroy its host's nerves. Torchin would come in some mornings and find shore crabs on their backs, still breathing but completely paralyzed. Within a few days the infected shore crabs died, and *Sacculina* died with them.

The biologists had come up hard against the trouble with parasites: their flexibility. Parasites may become specialists on a single host thanks to their evolutionary arms race. But that doesn't always mean that a parasite can't use the same tricks to infect another species. If it should come across a new host with a similar physiology and a similar way of life, it may be able to eke out an existence in it. The parasite may simply never get a chance to try out that new host because of its ecology: if a species of tapeworm lives in a stingray in the Amazon, it probably won't get a chance to try out stingrays in New Guinea. But sometimes parasites do get a chance—when, for example, continents slam together and animals on one of them colonize the other. That, in fact, seems to be how parasites survive through mass extinctions that claim so many of their hosts. They just jump from one host to a new one.

And so parasites carelessly introduced to new habitats can cause disasters, for all the reasons that make them so impressive when they work well. They have a sophisticated set of tactics they can use against their hosts, and they can fine-tune them through evolution to take on new hosts and new defenses. And once they get into a new habitat, there's no way to get them back out. It is a one-way experiment.

The halt of the cassava mealybug may be a great success story, but there are stories of spectacular failure as well. The forests of Hawaii represent one. They're filled with alien parasites brought there to destroy insect pests. Parasitic flies, for example, were brought in to keep down a species of stinkbug. But the fly could also live inside the koa bug, a big, showy native insect, and now the koa bug has almost disappeared. Parasitic wasps were brought in to control moths that attacked crops, and they also spread to many native species. Before the parasites came, the moths of Hawaii went through huge annual explosions; at their peak, their droppings falling from the trees sounded like a hailstorm. Birds would feast on their caterpillers and feed them to their young. But since the introduction of parasites, many native moths have managed to break out only once every decade or two. The forest birds of Hawaii are declining, and biologists suspect that the death of the moths may be partially to blame, because they can't feed the birds. And without birds to pollinate the trees and disperse their seeds, the forests themselves may also be suffering.

Hawaii's plight is the best documented of biological control's failure because it's a set of small, biologically distinct islands. But critics suspect that there are many other stories waiting to be told. In the United States, for example, over thirty different parasites were introduced during this century to kill gypsy moths. None of them worked well, and some of them have been destroying the exquisite giant silk moths, threatening them with extinction.

These disasters have made biologists like Lafferty and Kuris much more careful about using parasites. That was why they had set up such a long, tedious test of *Sacculina* in the first place. After the shore crabs started dying, they repeated their tests on Dungeness crabs. They got the same results: paralysis followed by death. "If I were to be responsible for the destruction of the Dungeness crab," Kuris said, "my name would be mud. I would be like the guy who introduced the killer bees. The poor man has lived a life of public self-flagellation for forty years. Do I care about the native shore crabs? Sure I do. I yield to nobody on values on this."

Lafferty broke the bad news to his colleagues in the fall of 1999. By then, the green crab had been spotted as far north as

How to Live in a Parasitic World

British Columbia, over a thousand miles from its landing point in San Francisco. Lafferty e-mailed me as well, and I immediately called him. I asked him if he was disappointed. "Well, as a scientist, you're never supposed to be disappointed," he said. "The truth exists, and you don't have any control over what's reality."

But it was frustrating to watch the green crab keep spreading. "My gut feeling is that if you released these things on the West Coast, chances are they wouldn't affect native crabs very much. All we found was that they have the potential to." Putting *Sacculina* larvae in a cup with a Dungeness crab isn't the same thing as putting them in the ocean. "It's got to ask these questions, like where is it likely to find its host crab."

Sacculina and its relatives use cues such as sunlight and chemicals given off by their hosts to position themselves where they're likely to bump into a green crab. Those cues might not let them bump into any other species. Lafferty told me about another experiment he had run that supported this idea. He got his hands on another species of parasitic barnacle that is related to *Sacculina* and lives in the Pacific sheep crab. He then gathered California shore crabs that live in the same range as the sheep crab, but which have never been found carrying a parasitic barnacle of their own. When he exposed the shore crab to the parasite, he had no trouble infecting it. Something must be preventing the parasite from infecting the crab in the wild.

But if you're trying to use parasites in the ocean as a biological control for the first time in history, you want to be utterly sure of yourself. I asked Lafferty if he had any other ideas for stopping the green crabs. "I don't think we should sit back and watch the massacre," he said. He started telling me about another parasite of green crabs called *Portunion conformis*. It's an isopod, a relative of pill bugs, and it has independently evolved a *Sacculina*-like existence of its own in green crabs. It invades a crab as a microscopic larva and then destroys its host's gonads, taking their place. Eventually it fills up a fair part of the crab's body, making up a fifth of its weight. By destroying the crab's gonads, it castrates its host, and like *Sacculina*, it feminizes male crabs. No one has ever cultured *Portunion* in a lab, but Lafferty wants to try. And then he wants to run the same tests on these parasites that *Sacculina* failed.

"They're absolutely beautiful parasites," Lafferty said. He had me picture a big, opaque pouch with a mouth at one end, carrying a collection of golden eggs inside. "It's hard to describe them. They look like—God, they don't look like anything you could ever imagine." Parasites may be frustrating to work with sometimes, but for a parasitologist, there's always a consolation in their beauty.

•   •   •

Herren and Lafferty work on the tattered edge of nature, the cassava fields and oyster banks where humans have transformed wilderness into a new sort of patchwork, where alien species can move thousands of miles in a matter of weeks, where the best-suited species is often the one that can thrive on perpetual chaos. Parasites may be able to soften the blow that we inflict in places like these if we respect their evolutionary power. But I also wondered about those parts of the world still left relatively untouched, and whether parasites might help keep them intact.

That was how I ended up in a Costa Rican jungle hunting frogs with Daniel Brooks. We were wandering around inside the Area de Conservación Guanacaste, a 220,000-acre reserve of dry forests, rain forests, and cloud forests, stretching from Pacific beaches to the tops of volcanoes. Twenty years ago, the forests of Guanacaste were disappearing as ranchers were cutting down trees to clear fields for their cattle, despite the fact that ranching was becoming less and less profitable. A biologist working in the area, a grizzled man named Daniel Janzen, decided to take advantage of the times. He set up a foundation that began buying up the ranches, and he hired the out-of-work cowboys to serve as "parataxonomists"—doing the work of documenting the diversity of Guanacaste by collecting species, dissecting them, and describing them. So the forest has not only been saved but expanded, and the people who live around it have a stake in protecting it. There are no fences around Guanacaste.

By the end of the 1990s, when I visited Guanacaste, Janzen was pretty much done with his reserve building. He was spending more of his time on his true love, the butterflies of Costa Rica. When you enter his little house at the reserve headquarters, three rooms under a corrugated tin roof, you have to stoop below the dozens of plastic bags hanging from the beams, each with a caterpillar feeding on a leaf. "My goal is to find all the caterpillars before I'm buried in the mud here," Janzen said to me. Not only does Guanacaste contain a fair amount of pristine forest, but more important, in the future its forests will grow and turn into a self-sustaining ecosystem. "A thousand years from now, you come back and it'll still be there," he said.

One night Brooks and I burst into Janzen's house. That day we had done a lot of dissections and looked at a lot of parasites, and we had decided to take a drive to a bar half an hour away for a drink. Along the way, the headlights of Brooks's four-by-four lit up a furry body on the road. We stopped and backed up. It was a dead fox freshly killed, its tail still a delicate cloud of gray. It went into the back of the truck, and we headed back to Guanacaste. When we got to Janzen's house, Brooks pulled the fox out and carried it to the front door. He laid it on the concrete floor of Janzen's front room. The animal looked intact, but it had been hit so hard that its eyes bulged like domes out of its head. Janzen said, "Well, what do we have here?"

Janzen's wife, Winnie, wandered out from the back room to see what was going on. She had their pet porcupine, Espinita, on her shoulder, and it raised its quills in fear. "You've been learning too much from your cats," Winnie said to Brooks, "bringing gifts to people's doors."

It takes a strong friendship to flop a bloody fox on someone's floor, and Janzen and Brooks have shared exactly that kind of friendship since 1994. (Janzen even named a species of parasitic wasp that he discovered after Brooks.) They met as Janzen was looking for help to count every species in the reserve. No one had ever done something on this vast scale—Janzen estimates that there are 235,000 species in Guanacaste. But he dreamed of having a full inventory of species, which scientists could use as a sort of yellow pages to let them pick out species to study and to help them discover how biodiversity is created and maintained in tropical forests. As soon as Brooks heard of the project, he wanted in.

Brooks has been a parasitologist since the mid-1970s. It was he who figured out how to use the relationships of parasites to reconstruct the wanderings of their hosts millions of years ago. He began working with frogs in Kansas but spent most of his career working in Latin America, looking at the parasites of stingrays, alligators, and other animals. It is slow work, and usually a parasitologist can hope to discover only a sliver of parasite diversity. And that's why Brooks jumped at Janzen's idea. "As soon as I heard about what was going on here," says Brooks, "I turned over all my stingray stuff to my Ph.D. students. I realized this was the place I wanted to make the focus of my work." For once, in one place, parasitologists might be able to know all the parasites. Guanacaste would become, as Brooks says, "a known parasite universe."

Janzen was a little puzzled by Brooks when they first met, and I could see some of that bafflement in his face when Brooks laid the fox on his floor. How can someone get so thrilled by a corpse? But Brooks evangelized Janzen until he began to see the parasitic light. "This guy shows up, and my vision of a mouse is changed forever," Janzen told me. "Now I see it as a bag of tapeworms and nematodes. You have this happy mouse and you open him up and he's full of them."

After showing off our find, Brooks and I took the fox to his shed. Brooks switched on the fluorescent light, and moths swarmed in through the chicken wire. He laid the fox down in the freezer, alongside an ocelot and a tapir—other lucky finds that he was going to open up eventually.

We got our drink—Cuba Libre in a can—and when we were done, at about eleven, we drove back to the reserve. Brooks pulled up by the shed and switched the light back on. The best way to see parasites is to open up a fresh body. As a corpse decomposes, the parasites lose their bearings and drift away from their natural homes. Soon they start to die themselves, their bodies disintegrating. So Brooks pulled the fox out of the freezer and got out a pair of scissors.

The fox's inner ecology turned out to be pretty simple: it was loaded with hookworms, which had been gouging blood out of its bowels. "This guy had a screaming hookworm infection," Brooks said, pulling apart the fox's intestine under a microscope. What struck me most about the dissection was Brooks himself. He kept apologizing to the fox as he cut it open—"Sorry, sorry"—kept cursing its stupid death, kept complaining about how the collision had smashed its lungs. The other scientists who worked at Guanacaste looked on Brooks as something of a vampire, a scientist interested in the beautiful animals of the forest only if he could slit them open. But I had never seen someone mourn a dead animal so deeply.

Janzen's dream of a full inventory fell apart in 1996 during negotiations with the Costa Rican government. Janzen didn't like how the money for the project was going to be diverted from the central mission of counting species, so he decided he had to abandon it. "We shot the horse in the head," was how he put it to me. But Brooks was able to get enough money from the Canadian government to keep going with the parasites. He estimates that the nine hundred forty vertebrates of the reserve harbor eleven thousand parasite species (including only the parasitic animals and protozoa), most of which will be new to science. "It's going to take the rest of my career to do this inventory," Brooks said. I wondered why he was planning to put himself through so much pain.

Over the course of the next day, I put the question to him a few times and got a new answer each time. Biodiversity is a staggering thing in a tropical forest such as Guanacaste, but you can't see most of it without the aid of a scalpel. "There are undoubtedly more species of parasites than free-living organisms," says Brooks. "When you preserve a species of deer, you're preserving twenty species of parasites from four kingdoms."

If that's not enough, you can justify the project out of enlightened selfishness. Most medicines trace their genealogy to some natural compound in some organism, be it penicillin from a fungus or digitalis from foxglove. Only in the past few years have scientists begun to work their way through the parasite's pharmocopeia. *Cordyceps*, a fungus that invades insects and sprouts flowerlike stalks out of its body, is the source of cyclosporin, an important antibiotic. Hookworms produce molecules that clasp perfectly with clotting factors in human blood, and biotechnology companies are putting them through trials as blood thinners for surgery. Ticks can also tamper with our blood to make their drinking easier, using chemicals that not only dissolve clots but reduce inflammation and kill bacteria that try to enter a wound. There are other parasitic tricks that still await an explanation. Blood flukes can steal substances out of our own blood to camouflage themselves from the immune system, but no one has figured out how they do it. If scientists did, they might be able to apply their discovery to transplanted organs. A doctor might be able to pump a patient's blood through a donor lung and essentially turn it into a gigantic protected fluke. That could spare patients from the dangers of immune-suppressing drugs. And these are only a few parasites; who knows what sorts of chemicals the millions of others have evolved?

Another reason for a parasite inventory came up when Brooks and I took a day off from dissections. We drove up the side of Volcan Cacao, thrashing in the back of a Land Cruiser on a road made from boulders. Much of the forest up the sides of the mountain had been cut down by ranchers, but conservationists had bought the land back and were waiting for the forests to grow back down the slopes. We stopped driving at the border of the forest and hiked in, instantly dunked in an ocean of trees, blue morpho butterflies bounding through the shade like fish swimming overhead. A thin rain worked its way down through the thick canopy as we walked over a creek. Brooks stopped to look upstream and down. "This place should be packed with frogs," he said. And there was nothing.

Beginning in the late 1980s, frogs began to disappear from the high elevations of Central America. On Cacao, not a single species of frog can be found. At first biologists had no idea what was causing the deaths; all they knew was that the corpses of frogs were piling up, untouched by birds. Only in 1999 did a biologist isolate what is probably the cause: a fungus brought down from the United States. Its spores travel through water until they meet the skin of a frog. Thereupon they dig into the animal and devour the keratin in its skin, releasing a toxin that quickly kills it. The only thing that keeps the fungus from killing every frog in Central America is the fact that it's adapted for cool climates, and it's too hot for the fungus to survive below a thousand meters.

By the time scientists had recognized the fungus, it was far too late to do anything. They could only watch the parasite bound southward from mountain to mountain. "We should have known about that fungus," says Brooks. "If we'd had an inventory of parasites of frogs, we might still have frogs on the mountaintops of Central America. We didn't know it was there." Humans have no special protection from parasites either, and they can come bounding out of disturbed rain forests. It won't be doctors who figure out where the Ebola virus comes from, but zoologists who can find the animal in the African rain forest that normally harbors it.

But Brooks doesn't look at his inventory simply as a catalog of death and destruction. It may be able to help scientists measure the ecological health of Guanacaste and other forests like it. An ecosystem is a bit like a person. In a healthy person, all the parts interact the way they should: the lungs take in oxygen and the stomach takes in food, the blood carries it all to the tissues, the kidneys flush out waste, and the brain ponders the world or what it wants for dinner. In a sick person, a few of the parts stop working, and their shutting down disrupts the person's whole body, sometimes forcing the rest of the parts to shut down as well. An ecosystem lasts for thousands or millions of years because it has parts that work together well: the worms aerate the soil, the fungus mingled with tree roots supplies them with nutrients and extracts carbohydrates in exchange, and so on. Water, minerals, carbon, and energy all circulate through the ecosystem like blood. And ecosystems, it turns out, can sicken. Introduce a parasite that kills koa bugs, and the damage can ripple out all the way to the trees in a forest.

Doctors don't wait until their patients are dead to declare that something's wrong with them. They look for early, easy-to-detect clues to trouble, even if they don't know yet what the trouble is. If a potentially fatal colony of bacteria have established themselves somewhere in a person's body, you don't have to actually track the microbes down—you can just check for a fever. Ecologists want something that can tell them that an ecosystem is sick before the damage has rippled out to all the strands of its web. They have been auditioning the species that make up ecosystems in the hopes of finding one that could act as a sort of body-temperature index. Some have been looking at ants and other insects, others at the songbirds that nest on forest floors. Many candidates fall short in one way or another. It's relatively easy to tell whether top predators such as wolves are declining, since they're relatively few and big. But by the time the effects of some environmental stress have surged all the way up the food chain to the wolf, the ecosystem is probably already too far gone to help.

How to Live in a Parasitic World

Some scientists, such as Brooks, think that parasites are a sign of ecological health, but not in the way most people would think. Until recently, most ecologists looked at parasites as nothing but a sign of environmental decline. If some pollutant wears down the immune systems of the members of an ecosystem, they become more susceptible to diseases. That does indeed seem to be true some of the time, but it's easy—and wrong—to make it a generalization. The idea echoes all the way back to Lankester: the rise of parasites as a sign of degenerate times. The frogs Brooks and I had collected in the lower forests were healthy and so abundant that they threw themselves across our path, and they were riddled with parasites. Parasites are actually a sign of an intact, unstressed ecosystem, and the opposite, as strange as it may sound, is true: if the parasites disappear from a habitat, it's probably in trouble.

As parasites travel through their life cycle they are often vulnerable to poisoning by pollution. A fluke, for example, hatches into a delicate form covered with hairlike cilia that swim in search of a snail; a couple of generations later, a cercaria emerges from the snail to find its mammal host. At both stages, the parasite depends on clean water to survive. That's the theory, at any rate, and there's some concrete evidence to show that it's correct. The rivers of Nova Scotia have become acidified as a result of air pollution from coal plants upwind. Canadian ecologists added lime to the headwaters of one badly hit river, neutralizing the acid there, and then came back in the following years to collect the eels. They then compared them with eels from an untreated river that eventually joined the limed one. The eels from the limed river carried inside them a much richer diversity of tapeworms, flukes, and other parasites. The ecologists then expanded their survey to the rivers along much of the coast of Nova Scotia, and found that the most badly affected waters had eels that were the most free of parasites.

Parasites work well as ecological sentinels for another reason: they sit at the top of many ecological webs. If you dump

241

nickel into a river, the little animals take up a little of it and don't suffer too badly, but as the nickel rises up the food web—as copepods are eaten by small fish, which are in turn eaten by big fish, which are in turn eaten by birds—the pollution focuses to higher and higher concentrations. But parasites, which prey on even the top predators, concentrate even more pollution in their bodies. Tapeworms may carry hundreds of times more lead or cadmium than the fish they travel inside, and thousands more than the surrounding water.

Unlike free-living organisms, a parasite wanders through the many levels of its ecosystem, and it can report on the damage it comes across in its travels. Throughout its life cycle, a parasite may need to move through many hosts, each of which occupies its own niche in the habitat. Flukes in the Carpinteria salt marsh have to live in snails, which depend on the algae on the mud banks; from there they find a fish, which must eat zooplankton to survive; and finally the parasite must find the gut of a healthy bird in which it can mature. If any of those hosts should disappear, the parasite will suffer. In 1997, Kevin Lafferty found that in the degraded part of the Carpinteria salt marsh, there are only half the species of parasites as in the healthy part, and only half the number of individual parasites. Parts of the marsh are now getting restored, and by 1999, the snails there had regained the levels of parasites found in the pristine marsh.

This is why Brooks is cutting open frogs in Costa Rica. "You've got this guy walking around with nine or ten parasites, healthy and happy. Once you know all the parasites in the frogs, suddenly if something's not there, something's wrong with the frogs or with an intermediate host. If you've lost a parasite, you have lost something in the fabric of the ecosystem." And once Brooks is done with his inventory, it may be possible to identify parasites by their eggs or larvae—and it won't be necessary to sacrifice any more hosts.

Parasites may not only mark good ecological health; they

may actually be vital for it. When ranchers overgraze their cattle and sheep on fragile grasslands, they can tip the ecology of the region over into a desert. As far as ecologists can tell, this move is pretty much irreversible, because the desert shrubs reorganize the soil in such a way that grasses can't penetrate back in. It is a difficult and politically volatile matter to decide just how much grazing should be allowed on a given patch of land. Ranchers usually dope up their livestock with medicine to kill as many intestinal worms as they can, but the parasites might be able to keep the livestock in a careful balance with the grass they depend on. The larvae of some species of parasitic worms get into livestock by attaching to the grass they eat. When a worm gets into the gut of a sheep, it matures and starts siphoning off some of the sheep's meals. Struggling with the effects of the worm, the sheep tends to live a shorter life and produce fewer lambs. In the end, the parasite shrinks the size of the herd.

Such ups and downs can alter an entire ecosystem. If a rancher is overgrazing his sheep on a semiarid grassland, the sheep may multiply and the plants will dwindle. At the same time the grazing changes the parasites: with more sheep available, they can breed in huge numbers, and they crowd on the dwindling blades of grass, making the probability that a sheep will become infected even higher. In other words, overgrazing automatically triggers an outbreak and scales back the herd, allowing the grass to recover. Soon the sheep population bounces back as well, but thanks to the management of the parasites, it never gets large enough to turn the grassland into desert. Rather than loading up their livestock with antiparasite drugs, and thereby ruining their grazing lands, ranchers may benefit by letting parasites keep the herd in check.

For now, though, the theory of parasitic stability remains mostly theory because scientists know so little about parasites in nature—which is another reason why Daniel Brooks is in Costa Rica. "People will be able to test their ideas on parasite stability because this won't be a parking lot in thirty years. Parasites may dampen oscillations, and if they are having an influence, you don't want to eradicate parasites."

To manage Guanacaste, in other words, you need to understand its parasites. "If we want to preserve a place like this," Brooks said, "we have to know what's going on microscopically. We need to figure out how to work with parasites. We need to figure out what organisms need and want, so we can use them in ways that don't terminate their existence."

The way Brooks was talking about us humans reminded me of the way parasites use their hosts—evolving a sense of what their hosts need and want, what they can and can't live without—so that they don't destroy themselves. In my travels for this book I often thought about the natural world as the sum of its parts. I would look down out of planes at the mud lakes of Sudan, the circuit-board housing tracts around Los Angeles, the disintegrating ranches and scraps of forest of Costa Rica and think about a concept, called Gaia, which some scientists embrace. They think of the biosphere—the rind of ocean, land, and air that's home to life—as a kind of superorganism. It has a metabolism of its own, which shuttles carbon and nitrogen and other elements around the world. The phosphorus that helps power the flash of a firefly ends up in the soil when the firefly dies, perhaps to be taken up by a tree and added to one of its leaves, dropping into a river and flowing to the sea, where photosynthesizing plankton take it up, only to be eaten by some grazing krill, which releases it into the ocean depths in its feces, only to be taken up by some bacterial scrounger, and cycled back up to the ocean's surface, before finally, many years later, ending up entombed in the sea floor. Like our own bodies, Gaia is held together and kept stable by its metabolism.

We humans exist within Gaia, and we depend on it for our survival. These days we live by using it up. We strip topsoil away with our farms without replacing it; we fish out the seas; we clear out forests. I thought about what Brooks had just said, about learning how to use nature without terminating it.

How to Live in a Parasitic World

"You talk as if we were a parasite," I said.

Brooks shrugged his shoulders. The idea was fine with him. "A parasite that has no self-regulation is going to put itself out of existence and may take its host with it," he said. "And the fact that most species on Earth are parasites tells us that hasn't happened a lot."

I chewed that over for a while. Here was a new meaning parasites could have for us—one that could take the place of Lankester's degenerates, Jewish tapeworms, and all the old myths of failed evolution. One that could be biologically faithful without turning life into a horror movie, without having parasites come bursting out of our ribs. It is we who are the parasites, and Earth the host. The metaphor may not be perfect, but it chimes well. We reroute the physiology of life to our own ends, mining fertilizer and blanketing farm fields with it, much as the wasp reroutes the physiology of its caterpillar to make the kind of foods it needs. We use up those resources and leave behind our waste, like *Plasmodium* turning a red blood cell into a garbage dump. If Gaia had an immune system, it might be disease and famine, which can keep an exploding species from taking over the world. But we have dodged these safeguards with medicines and clean toilets and other inventions, and they've allowed us to put billions of people on the planet.

There's no shame in being a parasite. We join a venerable guild that has been on this planet since its infancy and has become the most successful form of life on the planet. But we are clumsy in the parasitic way of life. Parasites can alter their hosts with great precision and change them for particular purposes: to take them back to their ancestral home in a stream, to move on to their adulthood inside a tern. But they are expert at causing only the harm that's necessary, because evolution has taught them that pointless harm will ultimately harm themselves. If we want to succeed as parasites, we need to learn from the masters.

Glossary

Antibody: A protein created by the immune system that can attach to antigens and neutralize them.

Antigen: A foreign substance that stimulates an immune response.

B cell: A type of immune cell that produces antibodies.

Blood fluke: One of several species of flukes that live in the bloodstream of vertebrates. The best studied are schistosomes, such as *Schistosoma mansoni*, which cause the disease schistisomiasis.

Chloroplast: A compartment in plants and algae where photosynthesis takes place. Originated as a free-living bacteria, which was engulfed by a eukaryote.

Complement: Blood-borne molecules that attack antigens, either on their own or in conjunction with antibodies.

Copepod: an aquatic crustacean that serves as an intermediate host to many parasites.

Cotesia congregata: A species of parasitic wasp that makes the tobacco hornworm its host.

Elephantiasis: A disease caused by filarial worms. The worms reside

Glossary

in the lymph channels, and the reaction of the immune system creates obstructions that trap lymph fluid in limbs or genitalia.

Flukes: Parasitic flatworms belonging to the class Trematoda.

Guinea worm: A parasitic nematode that lives in the abdomen of humans. After mating, the female emerges from her host's leg and releases larvae, which take up residence in a copepod.

Hookworm: A parasitic nematode that lives in the soil as a larva and as an adult in the human intestines. Consumes blood and causes anemia.

Macrophage: An immune cell that kills foreign organisms either by engulfing them or by releasing poisons.

Malaria: A disease characterized by high fever, caused by the protozoan *Plasmodium*.

Mast cell: Immune cell in linings of the intestines and nose; the cell can suddenly trigger allergic reactions.

Plasmodium: The protozoan that causes malaria.

River blindness: A disease caused by *Onchocerca volvulus*, a parasitic nematode. Blindness is caused by scarring triggered as the parasite crawls across the eyes.

Sacculina: A parasitic barnacle that lives in crabs.

Schistosomiasis: Also known as bilharzia. Disease caused by schistosomes, blood flukes that live in snails and humans. Its most serious symptom is liver damage caused by the immune system's reaction to schistosome eggs.

Sleeping sickness: Disease caused by the protozoan *Trypanosoma brucei*, and transmitted by the tsetse fly. Causes disorientation and coma. Fatal if not treated.

T cell: Immune cell that can recognize specific antigens. Killer T cells destroy cells infected with viruses and other pathogens. Inflammatory T cells organize attacks by macrophages. Helper T cells work with B cells to produce antibodies.

Glossary

Toxoplasma gondii: Protozoan that normally makes cats and their prey its hosts. Usually harmless in humans, except for pregnant women and people with compromised immune systems.

Trichinella: Parasitic nematode that lives in muscle cells.

Trypanosomes: Parasitic protozoa belonging to the genus *Trypanosoma.* Cause sleeping sickness (*T. brucei*), Chagas disease (*T. cruzi*), and other diseases.

Notes

Prologue: A Vein Is a River

xiv *"Trypanosoma brucei* has many enchanting features . . .": Borst et al., 1997, p. 121.

xvi Over 1.4 billion people carry the snakelike roundworm: These statistics come from Crompton, 1999.

1 Nature's Criminals

1 "Nature is not without a parallel . . .": Brown, 1898.

2 Eventually the parasite became a standard character: Damon, 1997.

2 Aristotle, for instance, recognized creatures: Grove, 1990.

2 two serpents wound around a staff: Roberts and Janovy, 2000.

3 "The substance in question cannot be a worm . . .": Quoted in Grove, 1990, p. 121.

4 "Some shoot forth horns . . .": Quoted in Wilson, 1995, p. 160.

5 The mysterious nature of parasites: See Farley, 1972.

6 "arches over them like a small, closely shut watch glass.": Quoted from Steenstrup 1845, pp. 57–58.

Notes

7   "An animal bears young . . .": Quoted from Steenstrup, 1845, p. 132.

8   "It would be contrary to the wise arrangement of Nature . . .": Quoted in Farley, 1972, p.120. For more details on the discovery of tapeworm life cycles, see also Grove, 1990, and Foster, 1965.

11   By 1900, bacteria were rarely called parasites anymore: Worboys, 1996.

11   When Leeuwenhoek had looked at his own feces: Roberts and Janovy, 2000.

13   When Napoleon took his army to Egypt: Nelson, 1990.

14   in the words of one scientist at the time, "medical zoology.": Worboys, 1983.

14   "It is derogatory that the Creator . . .": Quoted in Desmond and Moore, 1991, p. 293.

15   "I cannot persuade myself that a beneficent and omnipotent God . . .": Quoted in Desmond and Moore, 1991, p. 479.

15   To their mind, orthogenesis brought a purpose: Bowler, 1983.

15   One influential voice for orthogenesis: Lester, 1995.

16   "the jack-in-office, the pompous official . . .": Quoted in Lester, 1995, p. 59.

16   For biologists of Lankester's day: Cox, 1994.

17   "Let the parasitic life once be secured . . .": Lankester, 1890, p. 27.

17   Drummond declared that parasitism "is one of the gravest crimes . . .": Quoted from Drummond, 1883, p. 319.

18   "All those indiviudals who have secured a hasty wealth . . .": Quoted from Drummond, 1883, p. 350.

18   "In the struggle for daily bread . . .": Quoted from Hitler, 1971, p. 285.

18   "only and always a parasite in the body of other peoples. . . .": Quoted from Hitler, 1971, p. 304.

19   To Marx and Lenin: See Brennan, 1995.

19   "With the refinement of innate cruelty . . .": Quoted from Brown, 1898, pp. 162–163.

19   "Freedom, bondage, and the welfare state": Stunkard, 1955.

21   "When we use the terms 'higher and lower' . . .": Quoted from Lorenz, 1989, p. 41.

Notes

22   "A retrogression of specific human characteristics . . .": Lorenz, 1989, p. 45.

22   "I believe that I have given . . .": Quoted from Steenstrup, 1845, p. 8.

2 *Terra Incognita*

24   Consider the blood fluke *Schistosoma mansoni*: This description is drawn mainly from Basch, 1991.

27   This tiny nematode comes our way: Campbell, 1983.

29   Sukhdeo ignored the advice: Sukhdeo summarizes his work in Sukhdeo, 1997.

31   In tropical countries, between 30 and 90 percent of cattle carry them: Spithill and Dalton, 1998.

36   Each of these copepods looks so different: For an overview of parasitic copepods, see Benz, in preparation.

36   As they feed, tapeworms grow at a spectacular rate: Roberts and Janovy, 2000.

37   When we eat, peristalsis immediately ripples through our intestines: See Sukhdeo, 1997.

37   The intestines are also home to hookworms: See Hotez et al., 1995; Hotez and Prichard, 1995.

38   A biotechnology company has isolated these molecules: For information on the company's work, see its web site: www.corvas.com.

40   To do so, they set down hooks on the vessel wall: Naitza et al., 1998.

41   Fifteen seconds after the blast: Only one species of *Plasmodium* invades red blood cells this way: *P. falciparum*, which causes the most dangerous kind of malaria.

41   The core of hemoglobin: Ginsburg et al., 1999.

42   In other words, *Plasmodium* has to transform these mere corpuscles: This description of how *Plasmodium* invades and rebuilds blood cells is drawn from Foley and Tilley, 1995, 1998; Sinden 1985.

42   In either case, the parasitized red blood cell can start dragging: Lauer et al., 1997.

Notes

43   *Trichinella* is also a biological renovator: See Capo et al., 1998;
     Despommier, 1990; Polvere et al., 1997.

44   Plants are even hosts to parasitic plants: See Press and Graves,
     1995; Stewart and Press, 1990.

45   But many plant-eating insects spend: Thompson, 1994.

45   nematodes that live in plant roots: For reviews of root nematodes,
     see Bird, 1996; Niebel, et al.; 1994.

47   Bigger hosts tend to have more species of parasites in them: Poulin,
     1995.

47   On the gills of a single fish: Rhode, 1994. For other examples of
     parasite niches, see Roberts and Janovy, 2000; Kennedy and Gue-
     gan, 1996.

47   When parasitologists crack open the shells of snails: Kuris and Laf-
     ferty, 1994.

48   The wasp *Copidosoma floridanum*: Strand and Grbic, 1997.

50   The adult filarial worms live in the lymph channels: Roberts and
     Janovy, 2000.

50   The fleas on a female rabbit's skin: Hart, 1994.

51   Dig a few feet down into the hard summer dirt: For details of *Pseu-
     dodiplorchis*, see Tinsley, 1990; Tinsley, 1995, and the references
     therein.

3 *The Thirty Years' War*

55   A man came one day to the Royal Perth Hospital: Harris et al., 1984.

57   But here, at any rate, is a brief survey: Janeway and Travers, 1994.

60   In September 1909, a strong young man: Ross and Thomson, 1910.

62   "a struggle between the defensive powers of the infected body . . .":
     Quoted from Ross and Thomson, 1910, p. 408.

62   They play an exhausting game of bait-and-switch: See Barry, 1997;
     Borst et al., 1997.

64   Because these latches can be recognized by the immune system:
     Borst et al., 1995.

65   Each species causes a disease of its own: Bloom, 1979.

Notes

65 *Leishmania* doesn't have to muscle its way: For details of *Leishmania*'s invasion, see Bogdan and Rollinghoff, 1999; Locksley and Reiner, 1995.

67 Few people know about *Toxoplasma:* For *Toxoplasma*'s evasions, see Sher, 1995.

70 One remarkable example is the tapeworm: White et al., 1997.

72 You can see their disguise at work in a simple experiment: Damian, 1987.

73 a paradox on the shores of Lake Victoria: Karanja et al., 1997.

74 Under the spell of the eggs: Leptak and McKerrow, 1997.

76 The parasite survives thanks to millions of viruses: For reviews of *Cotesia congregata* and its viruses, see Beckage, 1997, 1998; Dushay and Beckage, 1993; Lavine and Beckage, 1996.

4 A Precise Horror

79 biologists of his day just didn't know much: My description of *Sacculina* is drawn from Collis and Walker, 1994; DeVries et al., 1989; Gilbert et al., 1997; Glenner and Høeg, 1995; Glenner et al., 1989; Glenner et al., 2000; Hartnoll, 1967; Høeg, 1985a, 1985b, 1987, 1992, 1995; Lutzen and Høeg, 1995; O'Brien and Van Wyk, 1986; O'Brien and Skinner, 1990; Raibaut and Trilles, 1993.

82 This puppetry takes different forms: For general reviews of host manipulation, see Moore, 1995; Moore and Gotelli, 1996; Poulin, 1994.

82 Rather than just passively soak up the food: Thompson, 1993.

83 A fungus called *Puccinia:* Roy, 1993.

84 The wasps seem to be responsible for the anorexia: Adamo, 1998.

84 Another species of wasp goes even further: Brodeur and Vet, 1994.

84 There are parasitic nematodes: Vance, 1996.

86 A fungus that lives inside house flies: Krasnoff et al., 1995.

87 Along the coasts of Delaware lives a fluke: Curtis, 1987, 1990.

87 Known as *Dicrocoelium dendriticum:* Roberts and Janovy, 2000.

Notes

88 The guinea worm spends its early life: Roberts and Janovy, 2000.

90 When a mosquito lands on your arm: For the challenges mosquitoes face, and the way *Plasmodium* manipulates them, see Day and Edman, 1983; James and Rossignol, 1991; Koella, 1999; Koella et al., 1998b; Ribeiro, 1995.

90 A mosquito with ookinetes in it: Anderson et al., 1999.

91 A fluke called *Leucochloridium:* Roberts and Janovy, 2000.

91 Some species of tapeworms live in the guts of: LoBue and Bell, 1993.

91 They can also alter the behavior: Tierney et al., 1993.

92 A small crustacean named: Helluy and Holmes, 1989.

92 *Toxoplasma*, the protozoan lodged: Berdoy et al., 2000.

95 Moore built chambers out of Pyrex pie plates: Moore, 1983.

97 Their hunger pushes the sticklebacks to take more risks: Milinski, 1990.

97 Biologists have pulled out the neurons of *Gammarus:* Helluy and Holmes, 1989; Maynard et al., 1996.

99 Beetles are lured to egg-bearing droppings: Evans et al., 1992.

99 If you trap the fragrance of infected dung: Evans et al., 1998.

99 the tapeworm then uses more chemicals: Hurd, 1998; Webb and Hurd, 1999.

100 Put it on a pile of flour: Robb and Reid, 1996.

100 But once the tapeworm reaches maturity: Blankespoor et al., 1997.

104 oceans are swarming with viruses: Fuhrman, 1999.

104 For decades, ecologists who worked on the Serengeti: Dobson, 1995.

106 In fact, if you were to get rid of the fluke: Lafferty, 1993a.

108 The results were even more stark: Lafferty describes his experiments in Lafferty, 1997a; Lafferty and Morris, 1996.

109 But why would birds: Lafferty models the trade-offs for hosts like these birds in Lafferty, 1992.

109 ecologist Greta Aeby has been scuba diving: Aeby, 1992, 1998.

111 they've contained fifteen quarts of fluid: Roberts and Janovy, 2000.

111 The thinning of the herd is an illusion: Messier et al., 1989; Rau and Caron, 1979.

115 It makes the corpse of its host a sexual magnet: Møller, 1993.

Notes

115 "I wonder why the titans . . .": Quoted from Heinlein, 1990, p. 205.

116 delusional parasitosis: Wykoff, 1987.

5 The Great Step Inward

122 The closest match he found: For the discovery of the apicoplast and its relationship with chloroplasts, see Kohler et al., 1997, and the references therein.

125 the eukaryotes with their DNA: Some of the most primitive eukaryotes such as *Giardia* are missing mitochondria, but recent gene sequencing has suggested that they originally had the organelle and lost it later in their evolution. (See, for instance, Hashimoto et al., 1998.) These results point to the first eukaryotes as having mitochondria.

125 the dawn of the age of eukaryotes: Knoll and Carroll, 1999.

126 Parasitism is any arrangement: Dawkins, 1982.

127 genetic parasites: Sherratt, 1995.

127 Some of them steal genes from their host: Xiong and Eickbush, 1990.

127 How is it, for instance, that a freshwater: Robertson, 1997.

128 Eventually the coalition of genes got organized: For this promiscuous vision of the beginning of life, see Woese, 1998.

128 It was probably at this time that life began to diverge: Katz, 1998.

129 If the cost of trying to fight off the invasion: Law, 1998.

129 But biologists now recognize: Doolittle, 2000.

129 Among the fully sequenced species is *Rickettsia*: Muller and Martin, 1999.

130 This billion-year-old drama: Roos et al., 1999.

131 David Roos and his colleagues have speculated: Waller et al., 1998.

131 It wasn't until about 700 million years ago: Knoll and Carroll, 1999.

131 Soon afterward, animals came on shore: Zimmer, 1998.

131 at least fifty times other lineages of animals followed suit: Poulin, 1998.

133 Attacking people is not how the candiru makes a living: Kelley and Atz, 1964.

Notes

133 There you find nests of the ant *Tetramorium:* Holldobler and Wilson, 1990.

134 Some butterflies, for example, can trick ants: Akino et al., 1999.

135 A single cuckoo starts life much bigger than a warbler: Kilner et al., 1999.

136 The fetus faces the same troubles: Villereal, 1997.

136 This conflict plays out: Pennisi, 1998.

141 Parasites, in other words, have evolutionary stories: Brooks explains how to use this method in Brooks and McLennan, 1993.

142 Tapeworms probably first evolved: Hoberg et al., 1999a.

144 The thorn forests of Bolivia are home to marsupials: For their link to Australian mammals and parasites, see Gardner and Campbell, 1992.

145 Pterosaurs began sharing the sky with birds: Hoberg et al., 1999b.

146 The scenario that reconciles these facts best: Brooks, 1992.

147 The closest relatives to human tapeworms: Hoberg et al., 2000.

148 Suzanne Sukhdeo has sorted through the close relatives: Sukhdeo et al., 1997.

148 Parasitologists have compared species of nematodes: Read and Skorping, 1995.

149 "boring by-product.": Dawkins, 1990.

149 These are galls: For an overview of galls, see Shorthouse and Rohfritsch, 1992.

149 Warren Abrahamson of Bucknell University: Abrahamson, 1997.

151 A German evolutionary biologist named Dieter Ebert: Ebert, 1994.

152 And quite often, that optimal virulence: Ebert and Herre, 1996.

154 The biologist Edward Herre studied fig wasps: Herre, 1993.

155 The laws of virulence are also built: Ewald, 1995.

6 Evolution from Within

157 "We behold the face of nature . . .": Quoted from Darwin, 1857, p. 116.

158 "Good, when young, bad for the past 33 years.": Quoted in Adler, 1997.

Notes

158 he had Chagas disease: Adler, 1989.

158 Chagas disease is caused by *Trypanosoma cruzi*: Bastien, 1998.

158 Ticks and lice may only live on their host's skin: Mooring and Hart, 1992.

159 This sort of monitoring still goes on today: Bingham, 1997.

160 A. R. Kraaijeveld of the Imperial College in England: Kraaijeveld et al., 1998.

163 In only fifty generations: Lively, 1996.

167 It didn't take Lively long to see a clear pattern: Lively, 1987.

167 In a single lake, they could see parasites: Fox et al., 1996.

168 In Nigeria there lives another snail: Schrag et al., 1994a, 1994b.

168 The most unexpected support for the Red Queen's effect: Gemmill et al., 1997.

169 It gets into the skin of the rat: Koga et al., 1999.

169 In other words, *Strongyloides* can complete its life cycle: Viney, 1999.

170 For five years he and another of his postdoctoral students: Dybdahl and Lively, 1998.

171 "I should advise you to walk the other way.": This parallel between science and literature was nicely observed in Lythgoe and Read, 1998.

172 Hamilton and Zuk gathered together reports: Hamilton and Zuk, 1982.

173 In many of the tests—especially the lab experiments: Clayton, 1991.

173 Zuk studied red jungle fowl from Southeast Asia: Zuk et al., 1995.

173 In a more elaborate study, Swedish scientists: Schantz et al., 1996.

173 That certainly seems to be what's going on with the fish: Taylor et al., 1998.

174 Immune studies give the Hamilton-Zuk hypothesis: See Møller, 1999.

175 Mice, for example, can smell the urine: Kavaliers and Colwell, 1995a, 1995b.

175 "The scent of a male mouse . . .": Penn and Potts, 1998.

175 Bees may be having so much sex: Baer and Schmid-Hempel, 1999.

176 Many insects are shaped expressly to fend off parasites: Gross, 1993.

176 Thousands of species of ants: Feener and Brown, 1997.

Notes

178 Mammals are continually assaulted by parasites: The effects of parasites on mammal herds can be found in Hart, 1994, 1997; Hart and Hart, 1994; Hart et al., 1992; Mooring and Hart, 1992.

179 the howler monkeys of Central America: Personal communication, Dr. Katherine Milton.

180 Consider leaf-rolling caterpillars: Caveney et al., 1998.

180 They keep their distance because the manure: Hart, 1997.

181 The odor is like perfume: DeMoraes et al., 1998.

182 Some will just stop eating: Kyriazakis et al., 1998.

182 The woolly bears, in other words: Karban and English-Loeb, 1997.

182 That still gives the snails a month: Minchella, 1985.

183 If a fluke gets into a snail that's still sexually immature: Lafferty, 1993b.

183 When the fruit flies of the Sonoran desert are attacked by parasites: Polak and Starmer, 1998.

183 Lizards are also tormented by mites of their own: Sorci and Clobert, 1995.

184 Worker bumblebees spend their days flying: Muller and Schmid-Hempel, 1993.

184 When a lungworm drops to the ground in the manure: Robinson, 1962.

186 A new species is born out of isolation: For an accessible overview of speciation, see Weiner, 1994.

186 A parasite that prefers many different hosts: Kawecki, 1998.

187 Lineages of parasites may be able to resist extinction: Bush and Kennedy, 1994.

187 This local struggle: Thompson, 1998.

187 And as these populations of hosts fight off: Thompson, 1994.

188 An interrupted gene may suddenly become able: MacDonald, 1995.

188 The genes that make the receptors: Roth and Craig, 1998.

188 And once a genetic parasite has established itself: DeBerardinis et al., 1998.

189 A bacterium called *Wolbachia*: See Hurst, 1993; Hurst et al., 1999; Werren, 1998.

Notes

7 *The Two-Legged Host*

192 It's been worked out best for *Trichinella*: Bell, 1998.

194 A blood fluke that swam from snails to rats: Despres et al., 1992.

194 The trypanosomes humans had left behind: Stevens and Gibson, 1999.

195 In those early days, parasites did best: Hill et al., 1994.

195 By spreading cats and rats around most of the world: Cox, 1994.

195 Along the Andes, the houses that Incas built: Bastien, 1998.

196 The mosquitoes that carry malaria: Bruce-Chwatt and de Zulueta, 1980.

196 One sort of mutation in the beta chain: Friedman and Trager, 1981.

197 Called ovalocytosis, this disorder: Jarolim et al., 1991; Schofield et al., 1992.

197 One of the few clear signs from antiquity: Senok et al., 1997.

198 And archaeologists in Israel have found bones: Hershokovitz and Edelson, 1991.

199 These mild cases of malaria immunize children: Miller, 1996.

199 In 1990, a biologist named Bobbi Low: Low, 1990.

199 The signs might not be visible either: Penn and Potts, 1998.

200 According to Robin Dunbar: Dunbar, 1996.

201 Sick chimps will sometimes search for strange food: Huffman, 1997.

203 "For the first time it is economically feasible for nations . . .": Quoted from Russell, 1955, p. 158.

203 There are more human intestinal worms than humans: These statistics come from Crompton, 1999.

203 Parasites like hookworm and whipworm: Nokes et al., 1992.

204 the disability-adjusted life year: Chan, 1997.

205 Consider the hideous case of guinea worms: Crompton, 1999; Peries and Cairncross, 1997.

205 Seventeen million people carry the parasite: Crompton, 1999.

206 If a person with river blindness takes the drug: Meredith and Dull, 1998.

207 When giant dams are built: Roberts and Janovy, 2000.

Notes

207 Chloroquine cures malaria: Ginsburg et al., 1999.

207 Now huge parts of the globe harbor malaria: The spread of resistant malaria is traced in Su et al., 1997.

209 The World Health Organization organized: Wilson and Coulson, 1998.

209 In 1998, human trials began: Shi et al., 1999.

211 These flukes can sense how many: Haseeb et al., 1998.

211 The vaccine could then conceivably cause more harm: Good et al., 1998.

211 Scientists have found that if they give an extra dose: Wynn et al., 1995.

212 If people were vaccinated so that their immune system: Haseeb et al., 1998.

212 One of the architects of the theory of virulence: Ewald, 1994.

214 In 1997, scientists at the University of Iowa: Newman, 1999.

214 Parasite-free living may also be responsible: Bell, 1996; Lynch et al., 1998.

8 *How to Live in a Parasitic World*

216 "Whenever the earth changed its form . . .": Quoted from Farley, 1977, p. 38.

218 Scientists first conceived of using parasites: Two reviews of biological control—both critical—are Howarth, 1991; and Simberloff and Stiling, 1996.

220 It may, for example, have saved much of Africa: The success of the cassava mealybug control program is reviewed in Herren and Neuenschwander, 1991.

231 The forests of Hawaii represent one: Howarth, 1991.

232 In the United States, for example: Boettner, 2000.

233 But if you're trying to use parasites in the ocean: Lafferty discusses the threat and promise of marine biological control in Lafferty and Kuris, 1996.

238 Ticks can also tamper with our blood: Durden and Keirans, 1996.

239 Only in 1999 did a biologist isolate: Morell, 1999.

Notes

240 An ecosystem is a bit like a person: For an introduction to ecosystem health, see Costanza et al., 1992.

241 Parasites are actually a sign: For an overview of parasites and ecological health, see Lafferty, 1997b.

241 Canadian ecologists added lime: Marcogliese and Cone, 1997.

242 Tapeworms may carry hundreds of times: Sures et al., 1999.

243 When ranchers overgraze their cattle and sheep: Grenfell, 1992.

244 a concept, called Gaia, which some scientists embrace: Volk, 1998.

Further Reading and Selected Bibliography

The parasitologist Robert Desowitz has written several popular books about parasites from a more medical perspective than this book uses (see Desowitz, 1983; Desowitz, 1991; Desowitz, 1997). For a wry, thorough textbook of parasitology (the kind that comes with an epigraph from Hunter Thompson), see Roberts and Janovy, 2000. A concise look at the evolution and ecology of parasites can be found in Poulin, 1998. Mark Ridley talks about the effects of sexual selection, including the Red Queen, in a book of the same name (Ridley, 1993).

Abrahamson, W. G. 1997. *Evolutionary ecology across three trophic levels: Goldenrods, gallmakers, and natural enemies*; Monographs in population biology. Princeton: Princeton University Press.

Adamo, S. A. 1998. Feeding suppression in the tobacco hornworm, *Manduca sexta*: costs and benefits to the parasitic wasps *Cotesia congregata*. *Canadian Journal of Zoology* 76:1634–1640.

Adler, C. 1989. Darwin's illness. *Israeli Journal of Medical Science* 25:218–221.

Adler, J. 1997. The dueling diagnoses of Darwin. *Journal of the American Medical Association* 277:1275.

Further Reading and Selected Bibliography

Aeby, G. S. 1992. The potential effect the ability of a coral intermediate host to regenerate has had on the evolution of its association of a marine parasite. *Proceedings of the Seventh International Coral Reef Symposium, Guam* 2:809–815.

———. 1998. A digenean metacercaria from the reef coral, *Porites compressa*, experimentally identified as *Podocotyloides stenometra*. *Journal of Parasitology* 84:1259–1261.

Akino, T., J. J. Knapp, J. A. Thomas, and G. W. Elmes. 1999. Chemical mimicry and host specificity in the butterfly *Maculinea rebeli*, a social parasite of *Myrmica* ant colonies. *Proceedings of the Royal Society of London B* 266:1419–1426.

Anderson, R. A., J. C. Koella, and H. Hurd. 1999. The effect of *Plasmodium yoelii nigeriensis* infection on the feeding persistence of *Anopheles stephensi* Liston throughout the sporogonic cycle. *Proceedings of the Royal Society of London B* 266:1729–1734.

Baer, B., and P. Schmid-Hempel. 1999. Experimental variation in polyandry affects parasite loads and fitness in a bumble-bee. *Nature* 397:151–154.

Barry, J. D. 1997. The biology of antigenic variation in African trypanosomes. In *Trypanosomiasis and leishmaniasis: Biology and control*, edited by G. Hide, J. C. Mottram, G. H. Coombs, and P. Holmes. New York: CAB International.

Basch, P. F. 1991. *Schistosomes: Development, reproduction, and host relations*. New York: Oxford University Press.

Bastien, J. W. 1998. *The kiss of death: Chagas' disease in the Americas*. Salt Lake City: University of Utah Press.

Beckage, N. E. 1997. The parasitic wasp's secret weapon. *Scientific American* 277(5):82–87.

———. 1998. Parasitoids and polydnaviruses. *BioScience* 48(4):305–311.

Bell, R. G. 1996. IgE allergies and helminth parasites: A new perspective on an old conundrum. *Immunology and Cell Biology* 74:337–345.

———. 1998. The generation and expression of immunity to *Trichinella spiralis* in laboratory rodents. *Advances in Parasitology* 41:159–217.

Benz, G. W. In preparation. *Evolutionary biology of siphonostome parasites of vertebrates (Siphonostomatoida: Copepoda)*.

Further Reading and Selected Bibliography

Berdoy, M., J. P. Webster, and D. W. Macdonald, 2000. Fatal attraction in rats infected with *Toxoplasma gondii*. *Proceedings of the Royal Society of London B*, in press.

Berenbaum, M. R., and A. R. Zangerl. 1998. Chemical phenotype matching between a plant and its insect herbivore. *Proceedings of the National Academy of Sciences* 95:13743–13478.

Bingham, P. M. 1997. Cosuppression comes to the animals. *Cell* 90:385–387.

Bird, D. M. 1996. Manipulation of host gene expression by root-knot nematodes. *Journal of Parasitology* 82:881–888.

Blankespoor, C. L., P. W. Pappas, and T. Eisner. 1997. Impairment of the chemical defence of the beetle, *Tenebrio molitor*, by metacestodes (cysticercoids) of the tapeworm, *Hymenolepis diminuta*. *Parasitology* 115:105–110.

Bloom, B. R. 1979. Games parasites play: How parasites evade immune surveillance. *Nature* 279:21–26.

Boettner, G. H., J. S. Elkington, and C. J. Boettner, 2000. Effect of a biological control introduction on three nontarget native species of saturniid moths. *Conservation Biology*, in press.

Bogdan, C., and M. Rollinghoff. 1999. How do protozoan parasites survive inside macrophages? *Parasitology Today* 15:22–28.

Borst, P., W. Bitter, P. Blundell, M. Cross, R. McCulloch, G. Rudenko, M. C. Taylor, and F. van Leeuwen. 1997. The expression sites for variant surface glycoproteins of *Trypanosoma brucei*. In *Trypanosomiasis and leishmaniasis: Biology and control*, edited by G. Hide, J. C. Mottram, G. H. Coombs, and P. Holmes. New York: CAB International.

———, R. McCulloch, F. Van Leeuwen, and G. Rudenko. 1995. Antigenic variation of malaria. *Cell* 82:1–4.

Bowler, P. J. 1983. *The eclipse of Darwinism*. Baltimore: Johns Hopkins University Press.

Brennan, W. 1995. *Dehumanizing the vulnerable: When word games take lives*. Chicago: Loyola University Press.

Brodeur, J., and L. E. M. Vet. 1994. Usurpation of host behavior by a parasitic wasp. *Animal Behavior* 48:187–192.

Brooks, D. R. 1992. Origins, diversification, and historical structure of the helminth fauna inhabiting neotropical freshwater stingrays (Potamotrygonidae). *Journal of Parasitology* 78(4):588–595.

————, and Deborah A. McLennan. 1993. *Parascript: Parasites and the language of evolution*. Washington: Smithsonian Institution Press.

Brown, J. 1898. *Parasitic wealth, or money reform: A manifesto to the people of the United States and to the workers of the world*. New York: Charles Kerr & Co.

Bruce-Chwatt, L. J., and J. de Zulueta. 1980. *The rise and fall of malaria in Europe*. Oxford: Oxford University Press.

Bush, A. O., and C. R. Kennedy. 1994. Host fragmentation and helminth parasites: Hedging your bets against extinction. *International Journal of Parasitology* 24:1333–1343.

Campbell, W. C., ed. 1983. *Trichinella and trichinosis*. New York: Plenum.

Capo, V. A., D. D. Despommier, and R. I. Polvere. 1998. *Trichinella spiralis*: Vascular endothelial growth factor is up-regulated within the nurse cell during the early phase of its formation. *Journal of Parasitology* 84(2):209–214.

Caveney, S., H. Mclean, and D. Surry. 1998. Faecal firing in a skipper caterpillar is pressure-driven. *Journal of Experimental Biology* 201:121–133.

Chan, M. S. 1997. The global burden of intestinal nematode infections—fifty years on. *Parasitology Today* 13(11):438–443.

Clayton, D. H. 1991. The influence of parasites on host sexual selection. *Parasitology Today* 7(12):329–334.

Collis, S. A., and G. Walker. 1994. The morphology of the naupilar stages of *Sacculina carcini* (Crustacea: Cirripedia: Rhizocephala). *Acta Zoologica* 75(4):297–303.

Costanza, R., B. G. Norton, and B. D. Haskell, eds. 1992. *Ecological health: New goals for environmental management*. Washington, D.C.: Island Press.

Cox, F. E. G. 1994. The evolutionary expansion of the sporozoa. *International Journal of Parasitology* 24:1301–1316.

Crompton, D. W. T. 1999. How much human helminthiasis is there in the world? *Journal of Parasitology* 85:397–403.

Further Reading and Selected Bibliography

Curtis, L. A. 1987. Vertical distribution of an estuarine snail altered by a parasite. *Nature* 235:1509–1511.

———. 1990. Parasitism and the movement of intertidal gastropod individuals. *Biological Bulletin* 179:105–112.

Damian, R. 1987. The exploitation of host immune responses by parasites. *Journal of Parasitology* 73(1):1–13.

Damon, C. 1997. *The mask of the parasite*. Ann Arbor: University of Michigan Press.

Darwin, C. 1857. *The origin of species*. London: John Murray.

Dawkins, R. 1982. *The extended phenotype: The gene as the unit of selection*. New York: W. H. Freeman.

———. 1990. Parasites, desiderata lists and the paradox of the organism. *Parasitology* 100:S63–S73.

Day, J. F., and J. D. Edman. 1983. Malaria renders mice susceptible to mosquito feeding when gametocytes are most infective. *Journal of Parasitology* 69:163–170.

DeBerardinis, R. J., J. L. Goodier, E. M. Ostertag, and H. H. Kazazian. 1998. Rapid amplification of a retrotransposon subfamily is evolving the mouse genome. *Nature Genetics* 20:288–290.

DeMoraes, C. M., W. J. Lewis, P. W. Pare, H. T. Alborn, and J. H. Tumlinson. 1998. Herbivore-infested plants selectively attract parasitoids. *Nature* 393:570–573.

Desmond, A., and J. Moore. 1991. *Darwin: The life of a tormented evolutionist*. New York: W. W. Norton.

Desowitz, R. S. 1983. *New Guinea tapeworms and Jewish grandmothers*. New York: W. W. Norton.

———. 1991. *The malaria capers*. New York: W. W. Norton.

———. 1997. *Who gave Pinta to the Santa Maria?* New York: W. W. Norton.

Despommier, D. D. 1990. *Trichinella spiralis:* The worm that would be virus. *Parasitology Today* 6(6):193–196.

Despres, L., D. Imbert-Establet, C. Combes, and F. Bonhomme. 1992. Molecular evidence linking hominid evolution to recent radiation of schistosomes (Platyhelminthes: Trematoda). *Molecular Phylogenetics and Evolution* 1:295–304.

DeVries, M. C., D. Rittschof, and R. B. Forward. 1989. Response by rhizocephalan-parasitized crabs to analogues of crab larval-release pheromones. *Journal of Crustacean Biology* 9:517–524.

Dobson, A. 1995. The ecology and epidemiology of rinderpest virus in Serengeti and Ngorongoro conservation area. In *Serengeti II: Dynamics, management, and conservation of an ecosystem*, edited by A. R. E. Sinclair and P. Arcese. Chicago: University of Chicago Press.

Doolittle, W. F. 2000. Uprooting the tree of life. *Scientific American* 282:90–95.

Drummond, H. 1883. *Natural law in the spiritual world*. London: Hodder and Stoughton.

Dunbar, R. 1996. *Grooming, gossip, and the evolution of language*. Cambridge, Massachusetts: Harvard University Press.

Durden, L. A., and J. E. Keirans. 1996. Host-parasite coextinction and the plight of tick conservation. *American Entomologist* (Summer):87–91.

Dushay, M. S., and N. E. Beckage. 1993. Dose-dependent separation of *Cotesia congregata*–associated polydnavirus effects on *Manduca sexta* larval development and immunity. *Journal of Insect Physiology* 39(12):1029–1040.

Dybdahl, M. F., and C. M. Lively. 1998. Host-parasite coevolution: Evidence for rare advantage and time-lagged selection in a natural population. *Evolution* 52(8):1057–1066.

Eberhard, W. G. 1990. Evolution in bacterial plasmids and levels of selection. *Quarterly Review of Biology* 65:3–22.

Ebert, D. 1994. Virulence and local adaptation of a horizontally transmitted parasite. *Science* 265:1084–1088.

———, and E. A. Herre. 1996. The evolution of parasitic diseases. *Parasitology Today* 12(3):96–101.

Evans, W. S., M. C. Hardy, R. Singh, G. E. Moodie, and J. J. Cote. 1992. Effects of the rat tapeworm *Hymenolepis diminuta* on the coprophagic activity of its intermediate host, *Tribolium confusum*. *Canadian Journal of Zoology* 70:2311–2314.

———, A. Wong, M. Hardy, R. W. Currie, and D. Vanderwel. 1998. Evidence that the factor used by the tapeworm, *Hymenolepis diminuta*, to direct the foraging of its intermediate host, *Tribolium confusum*, is a volatile attractant. *Journal of Parasitology* 84:1098–1101.

Further Reading and Selected Bibliography

Ewald, P. W. 1994. *Evolution of infectious disease.* Oxford: Oxford University Press.

———. 1995. The evolution of virulence: A unifying link between parasitology and ecology. *Journal of Parasitology* 81(5):659–669.

Farley, J. 1972. The spontaneous generation controversy (1700–1860): The origin of parasitic worms. *Journal of the History of Biology* 5(1):95–125.

———. 1977. *The spontaneous generation controversy from Descartes to Oparin.* Baltimore: Johns Hopkins University Press.

Feener, D. H., and B. V. Brown. 1997. Diptera as parasitoids. *Annual Review of Entomology* 42:73–97.

Foley, M., and L. Tilley. 1995. Home improvements: Malaria and the red blood cell. *Parasitology Today* 11(11):436–439.

———. 1998. Protein trafficking in malaria-infected erythrocytes. *International Journal of Parasitology* 28:1671–1680.

Foster, W. D. 1965. *A history of parasitology.* Edinburgh: E. & S. Livingstone.

Fox, J. A., M. F. Dybdahl, J. Jokela, and C. M. Lively. 1996. Genetic structure of coexisting sexual and clonal subpopulations in a freshwater snail (Potamopyrgus antipodarum). *Evolution* 50:1541–1548.

Friedman, M. J., and W. Trager. 1981. The biochemistry of resistance to malaria. *Scientific American* (March):156–164.

Fuhrman, J. A. 1999. Marine viruses and their biogeochemical and ecological effects. *Nature* 399:541–548.

Gardner, S. L., and M. L. Campbell. 1992. Parasites as probes for biodiversity. *Journal of Parasitology* 78(4):596–600.

Gemmill, A. W., M. E. Viney, and A. F. Read. 1997. Host immune status determines sexuality in a parasitic nematode. *Evolution* 51(2):393–401.

Gilbert, J., E. Mouchel-Vielh, and J. S. Deutsch. 1997. Engrailed duplication events during the evolution of barnacles. *Journal of Molecular Evolution* 44:585–594.

Ginsburg, H., S. A. Wray. and P. G. Bray. 1999. An integrated model of chloroquine action. *Parasitology Today* 15:357–360.

Glenner, H., and J. T. Høeg. 1995. A new motile, multicellular stage involved in host invasion by parasitic barnacles (Rhizocephala). *Nature* 377:147–150.

Further Reading and Selected Bibliography

————, A. Klysner, and B. Brodin Larsen. 1989. Cypris ultrastructure, metamorphosis and sex in seven families of parasitic barnacles (Crustacea: Cirripedia: Rhizocephala). *Acta Zoologica* 23:229–242.

————, J. T. Høeg, J. J. O'Brien, T. D. Sherman, 2000. Invasive vermigon stage in the parasitic barnacles *Loxothylacus texanus* and *L. panopaei* (Sacculinidae): closing of the rhizocephalan life-cycle. *Marine Biology*, in press.

Good, M. F., D. C. Kaslow, and L. H. Miller. 1998. Pathways and strategies for developing a malaria blood-stage vaccine. *Annual Review of Immunology* 16:57–87.

Grenfell, B. T. 1992. Parasitism and the dynamics of ungulate grazing systems. *The American Naturalist* 139:907–929.

Gross, P. 1993. Insect behavioral and morphological defenses against parasitoids. *Annual Review of Entomology* 38:251–73.

Grove, D. I. 1990. *A history of human helminthology.* London: CAB International.

Hamilton, W. D., and M. Zuk. 1982. Heritable true fitness and bright birds: A role for parasites? *Science* 218:384–387.

Harris, A. R. C., R. J. Russell, and A. D. Charters. 1984. A review of schistosomiasis in immigrants in Western Australia, demonstrating the unusual longevity of *Schistosoma mansoni. Proceedings of the Royal Society of Tropical Medicine and Hygiene* 78:385–388.

Hart, B. L. 1994. Behavioral defense against parasites: Interaction with parasite invasiveness. *Parasitology* 109:S139–S151.

————. 1997. Behavioral defense. In *Host-parasite evolution: General principles and avian models,* edited by D. H. Clayton and J. Moore. Oxford: Oxford University Press.

————, and L. A. Hart. 1994. Fly switching by Asian elephants: Tool use to control parasites. *Animal Behavior* 48:35–45.

————, L. A. Hart, M. S. Mooring, and R. Olubayo. 1992. Biological basis of grooming behavior in antelope: The body-size, vigilance and habitat principles. *Animal Behavior* 44:615–631.

Hartl, D. L., A. R. Lohe, and E. R. Lozovskaya. 1997. Modern thoughts on an ancient mariner: function, evolution, regulation. *Annual Review of Genetics* 31:337–358.

Further Reading and Selected Bibliography

Hartnoll, R. G. 1967. The effects of sacculinid parasites on two Jamaican crabs. *Journal of the Linnean Society (Zoology)* 46:275–295.

Harwood, C. L., I. S. Young, D. L. Lee, and J. D. Altringham. 1996. The effect of *Trichinella spiralis* infection on the mechanical properties of the mammalian diaphragm. *Parasitology* 113:535–543.

Haseeb, M. A., N. R. Bergquist, L. K. Eveland, and R. C. Eppard. 1998. Vaccination against schistosomiasis: Progress, prospects, and novel approaches. Paper read at Ninth International Congress of Parasitology, 1998, at Chiba, Japan.

Hashimoto, T., L. B. Sanchez, T. Shirakura, M. Muller, and M. Hasegawa. 1998. Secondary absence of mitochondria in *Giardia lamblia* and *Trichomonas vaginalis* revealed by valyl-tRNA synthetase phylogeny. *Proceedings of the National Academy of Sciences* 95:6860–6865.

Heinlein, R. A. 1990. *The puppetmasters.* New York: Ballantine Books.

Helluy, S., and J. C. Holmes. 1989. Serotonin, octopamine, and the clinging behavior induced by the parasite *Polymorphus paradoxus* (Acanthocephala) in *Gammarus lacustris* (Crustacea). *Canadian Journal of Zoology* 68:1214–1220.

Herre, E. A. 1993. Population structure and the evolution of virulence in nematode parasites of fig wasps. *Science* 259:1442–1445.

Herren, H. R., and P. Neuenschwander. 1991. Biological control of cassava pests in Africa. *Annual Review of Entomology* 36:257–283.

Hershokovitz, I., and G. Edelson. 1991. The first identified case of thalassemia? *Human Evolution* 6(1):49–54.

Hill, A. V. S., S. N. R. Yates, C. E. M. Allsopp, S. Gupta, S. C. Gilbert, A. Lalvani, M. Aidoo, M. Davenport, and M. Plebanski. 1994. Human leukocyte antigens and natural selection by malaria. *Proceedings of the Royal Society of London B* 346:378–385.

Hitler, A. 1971. *Mein Kampf.* Translated by Ralph Manheim. Boston: Houghton Mifflin Company.

Hoberg, E. P., S. L. Gardner, and R. A. Campbell. 1999a. Systematics of the Eucestoda: Advances toward a new phylogenetic paradigm, and observations on the early diversification of tapeworms and vertebrates. *Systematic Parasitology* 42:1–12.

Further Reading and Selected Bibliography

————, A. Jones, and R. A. Bray. 1999b. Phylogenetic analysis among the families of the Cyclophyllidea (Eucestoda) based on comparative morphology, with new hypotheses for co-evolution in vertebrates. *Systematic Parasitology* 42:51–73

————, R. L. Rausch., K. Eom, and S. L. Gardner. 2000. A phylogenetic hypothesis for species of the genus *Taenia* (Cyclophyllidea: Taeniidae). *Journal of Parasitology*, in press.

Høeg, J. T. 1985a. Cypris settlement, kentrogon formation and host invasion in the parasitic barnacle *Lernaeodiscus porcellanae* (Muller) (Crustacea: Cirripedia: Rhizocephala). *Acta Zoologica* 66:1–45.

————. 1985b. Male cypris settlement in *Clistrosaccus paguri* Lilljeborg (Crustacea: Cirripedia: Rhizocephala). *Journal of Experimental Marine Biology and Ecology* 89:221–235.

————.. 1987. Male cypris metamorphosis and a new male larval form, the trichogon, in the parasitic barnacle *Sacculina carcini* (Crustacea: Cirripida: Rhizocephala). *Philosophical Transactions of the Royal Society of London B* 317:47–63.

————. 1992. Rhizocephala. In *Microscopic anatomy of invertebrates.* New York: Wiley-Liss.

————. 1995. The biology and life cycle of the Rhizocephala (Cirripedia). *Journal of the Marine Biological Association of the United Kingdom* 75:517–550.

Hölldobler, B., and E. O. Wilson. 1990. *The ants.* Cambridge: Harvard University Press.

Hotez, P., J. Hawdon, and M. Cappello. 1995. Molecular mechanisms of invasion by *Ancyclostoma* hookworms. In *Molecular approaches to parasitology*, edited by J. C. Boothroyd and R. Komuniecki. New York: Wiley-Liss.

————, and D. J. Prichard. 1995. Hookworm infection. *Scientific American* 272(June):68–74.

Howarth, F. G. 1991. Environmental impacts of classical biological control. *Annual Review of Entomology* 36:485–509.

Huffman, M. A. 1997. Current evidence for self-medication in primates: A multidisciplinary perspective. *Yearbook of Physical Anthropology* 40:171–200.

Further Reading and Selected Bibliography

Hurd, H. 1998. Parasite manipulation of insect reproduction: Who benefits? *Parasitology* 116(Supplement):S13–21.

Hurst, G. D. D., F. M. Jiggins, and J. H. Graf von der Schulenbeurg. 1999. Male-killing *Wolbachia* in two species of insect. *Proceedings of the Royal Society of London B* 266:735–740.

Hurst, L. D. 1993. The incidences, mechanisms and evolution of cytoplasmic sex ratio distorters. *Biological Reviews of the Cambridge Philosophical Society* 68:121–193.

James, A. A., and P. A. Rossignol. 1991. Mosquito salivary glands: Parasitological and molecular aspects. *Parasitology Today* 7:267–271.

Janeway, C. A., and P. Travers. 1994. *Immunobiology: The immune system in health and disease*. London: Current Biology, Ltd.

Jarolim, P., J. Palek, D. Amato, K. Hassan, P. Sapak, G. T. Nurse, H. L. Rubin, S. Zhai, K. E. Sahr, and S. Liu. 1991. Deletion in erythrocyte band 3 gene in malaria-resistant Southeast Asian ovalocytosis. *Proceedings of the National Academy of Sciences* 88:11022–11026.

Karanja, D. M. S., D. G. Colley, B. L. Nahlen, J. H. Ouma, and W. E. Secor. 1997. Studies of schistosomiasis in western Kenya: I. Evidence for immune-facilitated excretion of schistosome eggs from patients with *Schistosoma mansoni* and human immunodeficiency virus coinfections. *American Journal of Tropical Medicine and Hygiene* 56(5):515–521.

Karban, R., and G. English-Loeb. 1997. Tachinid parasitoids affect host plant choice by caterpillars to increase caterpillar survival. *Ecology* 78:603–611.

Katz, L. A. 1998. Changing perspectives on the origin of eukaryotes. *Trends in Ecology and Evolution* 13:493–497.

Kavaliers, M., and D. Colwell. 1995a. Discrimination by female mice between the odors of parasitized and non-parasitized males. *Proceedings of the Royal Society of London B* 261:31–35.

———. 1995b. Odors of parasitized males induce aversive responses in female mice. *Animal Behavior* 50:1161–1169.

Kawecki, T. J. 1998. Red Queen meets Santa Rosalia: Arms races and the evolution of host specification in organisms with parasitic lifestyles. *American Naturalist* 152:635–651.

Further Reading and Selected Bibliography

Kazazian, H. H., and J. V. Moran. 1998. The impact of L1 retrotransposons on the human genome. *Nature Genetics* 19:19–24.

Kelley, W. E., and J. E. Atz. 1964. A pygidiid catfish that can suck blood from goldfish. *Copeia* (4):702–704.

Kennedy, C. R., and J. F. Guegan. 1996. The number of niches in intestinal helminth communities of *Anguilla anguilla*: Are there enough spaces for parasites? *Parasitology* 113:293–302.

Kilner, R. M., D. G. Noble, and N. B. Davies. 1999. Signals of need in parent-offspring communication and their exploitation by the common cuckoo. *Nature* 397:667–672.

Knoll, A. H., and S. B. Carroll. 1999. Early animal evolution: Emerging views from comparative biology and geology. *Science* 284:2129–2137.

Koella, J. C. 1999. An evolutionary view of the interactions between anopheline mosquitoes and malaria parasites. *Microbes and Infection* 1:303–308.

———, P. Agnew, and Y. Michalakis. 1998a. Coevolutionary interactions between host life histories and parasite life cycles. *Parasitology* 116:S47–S55.

———, F. L. Sorensen, and R. A. Anderson. 1998b. The malaria parasite, *Plasmodium falciparum*, increases the frequency of multiple feeding of its mosqutio vector, *Anopheles gambiae*. *Proceedings of the Royal Society of London B* 265:763–768.

Koga, M., A. Ning, and I. Tada. 1999. *Strongyloides ratti*: Migration study of third-stage larvae in rats by whole-body autoradiography after ^{35}S-methinonine labeling. *Journal of Parasitology* 85:405–409.

Kohler, S., C. F. Delwiche, P. W. Denny, L. G. Tilney, P. Webster, R. J. M. Wilson, J. D. Palmer, and D. S. Roos. 1997. A plastid of probably green algal origin in apicomplexan parasites. *Science* 275:1485–1488.

Kraaijeveld, A. R., J. J. M. Van Alphen, and H. C. J. Godfray. 1998. The coevolution of host resistance and parasitoid virulence. *Parasitology* 116(Supplement):S29–S45.

Krasnoff, S. B., D. W. Watson, D. M. Gibson, and E. C. Kwan. 1995. Behavioral effects of the entomopathogenic fungus *Entomophthora muscae* on its host *Musca domestica*: Postural changes in dying hosts and gated patterns of mortality. *Journal of Insect Physiology* 41(10):895–903.

Kuris, A. M., and K. D. Lafferty. 1994. Community structure: Larval trematodes in snail hosts. *Annual Review of Ecology and Systematics* 25:189–217.

Kyriazakis, I., B. J. Tolkamp, and M. R. Hutchings. 1998. Toward a functional explanation for the occurrence of anorexia during parasitic infections. *Animal Behavior* 56:265–274.

Lafferty, K. D. 1992. Foraging on prey that are modified by parasites. *American Naturalist* 140:854–867.

———. 1993a. Effects of parasitic castration on growth, reproduction and population dynamics of the marine snail *Cerithidea californica*. *Marine Ecology Progress Series* 96:229–237.

———. 1993b. The marine snail, *Cerithidea californica*, matures at smaller sizes where parasitism is high. *Oikos* 68:3–11.

———. 1997a. The ecology of parasites in a salt marsh ecosystem. In *Parasites and pathogens: Effects on host hormones and behavior*, edited by N. E. Beckage. New York: Chapman & Hall.

———. 1997b. Environmental parasitology: What can parasites tell us about human impacts on the environment? *Parasitology Today* 13:251–255.

———, and A. M. Kuris. 1996. Biological control of marine pests. *Ecology* 77(7):1989–2000.

———, and K. Morris. 1996. Altered behavior of parasitized killifish increases susceptibility to predation by bird final hosts. *Ecology* 77(5):1390–1397.

Lankester, E. R. 1890. Degeneration: A chapter in Darwinism. In *The advancement of science: Occasional essays and addresses*. London: Macmillan.

Lauer, S. A., P. K. Rathod, N. Ghori, and K. Haldar. 1997. A membrane network for nutrient import in red cells infected with the malaria parasite. *Science* 276:1122–1125.

Lavine, M. D., and N. E. Beckage. 1996. Temporal pattern of parasitism-induced immunosuppression in *Manduca sexta*. *Journal of Insect Physiology* 42(1):41–51.

Law, R. 1998. Symbiosis through exploitation and the merger of lineages in evolution. *Proceedings of the Royal Society of London B* 265:1245–1253.

Leptak, C. L., and J. H. McKerrow. 1997. Schistosome egg granulomas and hepatic expression of TNF-α are dependents on immune priming during parasite maturation. *Journal of Immunology* 158:301–307.

Lester, J. E. 1995. *E. Ray Lankester and the making of modern British biology*, edited by P. J. Bowler. London: British Society for the History of Science Monographs.

Lively, C. M. 1987. Evidence from a New Zealand snail for the maintenance of sex by parasitism. *Nature* 328:519–521.

———. 1996. Host-parasite coevolution and sex: Do interactions between biological enemies maintain genetic variation and cross-fertilization? *BioScience* 46(2):107–114.

LoBue, C. P., and M. A. Bell. 1993. Phenotypic manipulation by the cestode parasite *Schistocephalus solidus* of its intermediate host, *Gasterosteus aculeatus*, the threespine stickleback. *American Naturalist* 142:725–735.

Locksley, R. M., and S. L. Reiner. 1995. Murine leishmaniasis and the regulation of CD4+ cell development. In *Molecular approaches to parasitology*, edited by J. C. Boothroyd and R. Komuniecki. New York: Wiley-Liss.

Lorenz, K. 1989. *The waning of humaneness.* Translated by Robert Warren Kickert. Boston: Little, Brown and Co.

Low, B. S. 1990. Marriage systems and pathogen stress in human behavior. *American Zoologist* 30:325–339.

Lutzen, J., and J. T. Høeg. 1995. Spermatogonia implantation by antennular penetrations in the akentrodonid rhizocephalan *Diplothylacus sinensis* (Keppen, 1877) (*Crustacea: Cirripedia*). *Zoologischer Anzeiger* 234:201–207.

Lynch, N., I. A. Hagel, M. Palenque, and M. C. DiPrisco. 1998. Relationship between helminithic infection and IgE response in atopic and nonatopic children in a tropical environment. *Journal of Allergy and Clinical Immunology* 101:217–221.

Lythgoe, K. A., and A. F. Read. 1998. Catching the Red Queen? The advice of the rose. *Trends in Ecology and Evolution* 13:473–474.

MacDonald, J. F. 1995. Transposable elements: Possible catalysts of or-

Further Reading and Selected Bibliography

ganismic evolution. *Trends in Ecology and Evolution* 10:123–126.

Marcogliese, D. J., and D. K. Cone. 1997. Parasite communities as indicators of ecosystem stress. *Parasitologia* 39:227–232.

Margulis, L. 1998. *Symbiotic planet: A new look at evolution.* New York: Basic Books.

Maynard, B. J., L. DeMartini, and W. G. Wright. 1996. *Gammarus lacustris* harboring *Polymorphus paradoxus* show altered patterns of serotonin-like immunoreactivity. *Journal of Parasitology* 82:663–666.

Meredith, S. E. O., and H. B. Dull. 1998. Onchocerciasis: The first decade of mectizan treatment. *Parasitology Today* 14:472–473.

Messier, F., M. E. Rau, and M. A. McNeill. 1989. *Echinococcus granulosus* (Cestoda: Taeniidae) infections and moose-wolf population dynamics in southwestern Quebec. *Canadian Journal of Zoology* 67:216–219.

Milinski, M. 1990. Parasites and host decision-making. In *Parasitism and host behavior,* edited by C. J. Barnard and J. M. Behnke. London: Taylor & Francis.

Miller, L. H. 1996. Protective selective pressure. *Nature* 383:480–481.

Minchella, D. J. 1985. Host life-history variation in response to parasitism. *Parasitology* 90:205–216.

Møller, A. P. 1993. A fungus infecting domestic flies manipulates sexual behavior. *Behavioral Ecology and Sociobiology* 33:403–407.

————. 1999. Parasitism, host immune function, and sexual selection. *Quarterly Review of Biology* 74:3–20.

Moore, J. 1983. Response to an avian predator and its isopod prey to an acanthocephalan parasite. *Ecology* 64(5):1000–1015.

————. 1995. The behavior of parasitized animals. *BioScience* 45(2):89–98.

————, and N. J. Gotelli. 1996. Evolutionary patterns of altered behavior and susceptibility in parasitized hosts. *Evolution* 50(2):807–819.

Mooring, M. S., and B. L. Hart. 1992. Animal grouping for protection from parasites: selfish herd and encounter-dilution effects. *Behavior* 123:173-193.

Morell, V. 1999. Are pathogens felling frogs? *Science* 284:728–731.

Muller, C. B., and P. Schmid-Hempel. 1993. Exploitation of cold temperature as defense against parasitoids in bumblebees. *Nature* 363:65–67.

Muller, M., and W. Martin. 1999. The genome of *Rickettsia prowazekii* and some thoughts on the origin of mitochondria and hydrogenosomes. *BioEssays,* 21:377–381.

Naitza, S., F. Spano, K. J. H. Robson, and A. Cristranti. 1998. The thrombospondin-related protein family of apicomplexan parasites: The gears of the cell invasion machinery. *Parasitology Today* 14:479–484.

Nelson, G. S. 1990. Human behavior and the epidemiology of helminth infections: Cultural practices and microepidemiology. In *Parasitism and host behavior,* edited by C. J. Barnard and J. M. Behnke. London: Taylor & Francis.

Newman, A. 1999. In pursuit of an autoimmune worm cure. *The New York Times,* August 31, 1999, 5.

Niebel, A., G. Gheysen, and M. Van Montagu. 1994. Plant-cyst nematode and plant-root-knot nematode interactions. *Parasitology Today* 10(11):424–430.

Nokes, C., S. M. Grantham-McGregor, A. W. Sawyer, E. S. Cooper, B. A. Robinson, and D. A. P. Bundy. 1992. Moderate to heavy infections of *Trichuris trichiura* affect cognitive function in Jamaican school children. *Parasitology* 104:539–547.

O'Brien, J., and D. M. Skinner. 1990. Overriding of the molt-inducing stimulus of multiple limb autotomy in the mud crab *Rhithropanopeus harrisii* by parasitization with a rhizocephalan. *Journal of Crustacean Biology* 10(3):440–445.

———, and P. Van Wyk. 1986. Effects of crustacean parasitic castrators (epicaridean isopods and rhizocephalan barnacles) on growth of crustacean hosts. In *Crustacean Issues 3,* edited by A. Wenner.

Penn, D., and W. K. Potts. 1998. Chemical signals and parasite-mediated sexual selection. *Trends in Ecology and Evolution* 13:391–396.

Pennisi, E. 1998. A genomic battle of the sexes. *Science* 281:1984–1985.

Peries, H., and S. Cairncross. 1997. Global eradication of guinea worm. *Parasitology Today* 13(11):431–437.

Poirier, S. R., M. E. Rau, and X. Wang. 1995. Diel locomotor activity of deer mice (*Peromyscus maniculatus*) infected with *Trichinella nativa* or *Trichinella pseudospiralis*. *Canadian Journal of Zoology* 73:1323–1334.

Polak, M., and W. T. Starmer. 1998. Parasite-induced risk of mortality elevates reproductive effort in male *Drosophila*. *Proceedings of the Royal Society of London B* 265:2197–2201.

Polvere, R. I., C. A. Kabbash, V. A. Capo, I. Kadan, and D. D. Despommier. 1997. *Trichinella spiralis*: Synthesis of type IV and type VI collagen during nurse cell formation. *Experimental Parasitology* 86:191–199.

Poulin, R. 1994. The evolution of parasite manipulation of host behavior: A theoretical analysis. *Parasitology* 109:S109–S118.

———. 1995. Phylogeny, ecology, and the richness of parasite communities in vertebrates. *Ecological Monographs* 65(3):283–302.

———. 1998. *Evolutionary ecology of parasites: From individuals to communities*. London: Chapman & Hall.

Press, M. C., and J. D. Graves, eds. 1995. *Parasitic plants*. London: Chapman & Hall.

Raibaut, A., and J. P. Trilles. 1993. The sexuality of parasitic crustaceans. *Advances in Parasitology* 32:367–444.

Rau, M. E. 1983. The open-field behavior of mice infected with *Trichinella spiralis*. *Parasitology* 86:311–318.

———, and F. R. Caron. 1979. Parasite-induced susceptibility of moose to hunting. *Canadian Journal of Zoology* 57:2466–2468.

Read, A. F., and A. Skorping. 1995. The evolution of tissue migration by parasitic nematode larvae. *Parasitology* 111:359–371.

Ribeiro, J. M. C. 1995. Blood-feeding arthropods: Live syringes or invertebrate pharmacologists? *Infectious Agents and Disease* 4:143–152.

Ridley, M. 1993. *The Red Queen: Sex and the evolution of human nature*. New York: Penguin.

Robb, T., and M. L. Reid. 1996. Parasite-induced changes in the behavior of cestode-infected beetles: Adaptation or simple pathology? *Canadian Journal of Zoology* 1268–1274.

Roberts, L. S., and J. Janovy. 2000. *Gerald D. Schmidt and Larry Roberts' Foundation of Parasitology*. 6th ed. Dubuque, IA: McGraw-Hill.

Robertson, H. M. 1997. Multiple mariner transposons in flatworms

Further Reading and Selected Bibliography

and hydras are related to those of insects. *Journal of Heredity* 88:195–201.

Robinson, J. 1962. *Pilobolus* spp. and the translation of the infective larvae of *Dictyocaulus viviparus* from feces to pastures. *Nature* 193:353–354.

Rohde, K. 1994. Niche restriction in parasites: Proximate and ultimate causes. *Parasitology* 109:S69–S84.

Roos, D. S., M. J. Crawford, R. G. K. Donald, J. C. Kissinger, L. J. Klimczak, and B. Striepen. 1999. Origin, targeting, and function of the apicomplexan plastid. *Current Opinion in Microbiology* 2:426–432.

Ross, R., and D. Thomson. 1910. A case of sleeping sickness studied by precise enumerative methods: Further observations. *Proceedings of the Royal Society for Tropical Medicine* 82:395–409.

Roth, D. B., and N. L. Craig. 1998. VDJ recombination: A transposase goes to work. *Cell* 94:411–414.

Roy, B. A. 1993. Floral mimicry by a plant pathogen. *Nature* 362:56–58.

Russell, P. F. 1955. *Man's mastery of malaria*. Oxford: Oxford University Press.

Schantz, T. von, H. Wittzell, G. Goransson, M. Grahn, and K. Persson. 1996. MHC genotype and male ornamentation: Genetic evidence for the Hamilton-Zuk model. *Proceedings of the Royal Society of London B* 263:265–271.

Schofield, A. E., D. M. Reardon, and M. J. A. Tanner. 1992. Defective anion transport activity of the abnormal band 3 in hereditary ovalocytic red blood cells. *Nature* 355:836–838.

Schrag, S. J., A. Mooers, G. T. Ndifon, and A. F. Read. 1994a. Ecological correlates of male outcrossing ability in a simultaneous hermaphrodite snail. *American Naturalist* 143:636–655.

———, G. T. Ndifon, and A. F. Read. 1994b. Temperature-determined outcrossing ability in wild populations of a simultaneous hermaphrodite snail. *Ecology* 75:2066–2077.

Senok, A. C., K. Li, E. A. S. Nelson, L. M. Yu, L. P. Tian, and S. J. Oppenheimer. 1997. Invasion and growth of *Plasmodium falciparum* is inhibited in fractionated thalassaemic erythrocytes. *Transactions of the Royal Society of Tropical Medicine and Hygiene* 91:138–143.

Further Reading and Selected Bibliography

Sher, A. 1995. Regulation of cell-mediated immunity by parasites: The ups and downs of an important host adaptation. In *Molecular Approaches to Parasitology*, edited by J. C. Boothroyd and R. Komuniecki. New York: Wiley-Liss.

Sheratt, D. J., ed. 1995. *Mobile genetic elements*. New York: IRL Press.

Shi, Y. P., S. E. Hasnain, J. B. Sacci, B. P. Holloway, J. Fujioka, N. Kumar, R. Wohlheuter, S. L. Hoffman, W. E. Collins, and A. A. Lal. 1999. Immunogenicity and in vitro protective efficacy of a recombinant multistage *Plasmodium falciparum* candidate vaccine. *Proceedings of the National Academy of Sciences* 96:15–20.

Shorthouse, J. D., and O. Rohfritsch, eds. 1992. *Biology of Insect-Induced Galls*. Oxford: Oxford University Press.

Simberloff, D., and P. Stiling. 1996. How risky is biological control? *Ecology* 77 (7):1965–1974.

Sinden, R. E. 1985. A cell biologist's view of host cell recognition and invasion by malarial parasites. *Transactions of the Royal Society of Tropical Medicine and Hygiene* 79:598–605.

Sorci, G., and J. Clobert. 1995. Effects of maternal parasite load on offspring life-history traits in the common lizard (*Lacerta vivipara*). *Journal of Evolutionary Biology* 8:711–723.

Spithill, T. W., and J. P. Dalton. 1998. Progress in development of liver fluke vaccine. *Parasitology Today* 14 (6):224–228.

Steenstrup, J. 1845. *On the alternation of generations; or The propagation and development of animals through alternate generations: A peculiar form of fostering the young in the lower classes of animals*. London: The Ray Society.

Stevens, J. R., and W. Gibson. 1999. The molecular evolution of trypanosomes. *Parasitology Today* 15:432–436.

Stewart, G. R., and M. C. Press. 1990. The physiology and biochemistry of parasitic angiosperms. *Annual Review of Plant Physiology and Plant Molecular Biology* 41:127–151.

Strand, M. R., and M. Grbic. 1997. The life history and development of polyembryonic parasitoids. In *Parasites and Pathogens: Effects on Host Hormones and Behavior*, edited by N. E. Beckage. New York: Chapman & Hall.

Further Reading and Selected Bibliography

Stunkard, H. W. 1955. Freedom, bondage, and the welfare state. *Science* 121:811–816.

Su, X., L. A. Kirkman, H. Fujioka, and T. E. Wellems. 1997. Complex polymorphisms in an ~330 kDa protein are linked to chlorquine-resistant *P. falciparum* in Southeast Asia and Africa. *Cell* 91:593–603.

Sukhdeo, M. V. K. 1997. Earth's third environment: The worm's eye view. *BioScience* 47 (3):141–52.

Sukhdeo, S. C., M. V. K. Sukhdeo, M. B. Black, and R. C. Vrijenhoek. 1997. The evolution of tissue migration in parasitic nematodes (Nematoda: Strongylida) inferred from a protein-coding mitochondrial gene. *Biological Journal of the Linnean Society* 61:281–298.

Sures, B., R. Siddall, and H. Taraschewski. 1999. Parasites as accumulation indicators of heavy metal polluters. *Parasitology Today* 15:16–21.

Taylor, M. I., G. Turner, R. L. Robinson, and J. R. Stauffer. 1998. Sexual selection, parasites, and bower height skew in a bower-building cichlid fish. *Animal Behavior* 56:379–384.

Thompson, J. N. 1994. *The coevolutionary process.* Chicago: University of Chicago Press.

———. 1998. Rapid evolution as an ecological process. *Trends in Ecology and Evolution* 13:329–332.

Thompson, S. N. 1993. Redirection of host metabolism and effects on parasite nutrition. In *Parasites and Pathogens of Insects,* edited by N. E. Beckage, S. N. Thompson, and B. A. Federicic. New York: Academic Press.

Tierney, J. F., F. A. Huntingford, and D. W. T. Crompton. 1993. The relationship between infectivity of *Schistocephalus solidus* (Cestoda) and anti-predator behavior of its intermediate host, the three-spined stickleback, *Gasterosteus aculeatus. Animal Behavior* 46:603–605.

Tinsley, R. C. 1990. The influence of parasite infection on mating success in spadefoot toads, *Scaphiopus couchi. American Zoologist* 30:313–324.

———. 1995. Parasitic disease in amphibians: Control by the regulation of worm burdens. *Parasitology* 111:S153–S178.

Vance, S. A. 1996. Morphological and behavioral sex reversal in mermithid-infected mayflies. *Proceedings of the Royal Society of London B* 263:907–912.

Viney, M. E. 1999. Exploiting the life cycle of *Strongyloides ratti*. *Parasitology Today* 15:231–235.

Villereal, L. P. 1997. On viruses, sex, and motherhood. *Journal of Virology* 71:859–865.

Volk, T. 1998. *Gaia's body: Toward a physiology of earth*. New York: Copernicus.

Waller, R. F., P. J. Keeling, R. G. K. Donald, B. Stripen, E. Handman, N. Lang-Unnasch, A. F. Cowman, G. S. Besra, D. S. Roos, and G. I. McFadden. 1998. Nuclear-encoded proteins target to the plastid in *Toxoplasma gondii* and *Plasmodium falciparum*. *Proceedings of the National Academy of Sciences* 95:12352–12357.

Webb, T. J., and H. Hurd. 1999. Direct manipulation of insect reproduction by agents of parasite origin. *Proceedings of the Royal Society of London B* 266:1537–1541.

Weiner, J. 1994. *The beak of the finch: A story of evolution in our own time*. New York: Knopf.

Werren, J. H. 1998. *Wolbachia* and speciation. In *Endless forms: Species and Speciation*, edited by D. J. Howard. Oxford: Oxford University Press.

White, A. Clinton, P. Robinson, and R. Kuhn. 1997. *Taenia solium* cysticercosis: host-parasite interactions and the immune response. *Chemical Immunology* 1997:209–230.

Wilson, C. 1995. *The invisible world: Early modern philosophy and the invention of the microscope*. Princeton: Princeton University Press.

Wilson, R. A., and P. S. Coulson. 1998. Why don't we have a schistosomiasis vaccine? *Parasitology Today* 14(3):97–99.

Woese, C. R. 1998. The universal ancestor. *Proceedings of the National Academy of Sciences* 95:6854–6859.

Worboys, M. 1983. The emergence and early development of parasitology. In *Parasitology: A global perspective*, edited by K. S. Warren and J. Z. Bowers. New York: Springer-Verlag.

———. 1996. Germs, malaria and the invention of Mansonian tropical medicine: from "Diseases in the Tropics" to "Tropical Diseases." In

Further Reading and Selected Bibliography

Warm climates and western medicine: The emergence of tropical medicine, 1500–1900, edited by D. Arnold. Amsterdam: Rodopi.

Wykoff, R. F. 1987. Delusions of parasitosis: A review. *Review of Infectious Diseases* 9:433–437.

Wynn, T. A., A. W. Cheever, D. Jankovic, R. W. Poindexter, P. Caspar, F. A. Lewis, and Alan Sher. 1995. An IL-12-based vaccination method for preventing fibrosis induced by schistosome infection. *Nature* 376:594–596.

Xiong, Y., and T. Eickbush. 1990. Origin and evolution of retroelements based upon their reverse transcriptase sequences. *EMBO Journal* 9:3353–3362.

Zimmer, C. 1998. *At the water's edge: Macroevolution and the transformation of life.* New York: Free Press.

Zuk, M., T. S. Johnsen, and T. Maclarty. 1995. Endocrine-immune interactions, ornaments and mate choice in red junglefowl. *Proceedings of the Royal Society of London B* 260:205–210.

Acknowledgments

I researched this book by picking the brains of many scientists, either in person or via telephone lines and modems. Thanks go in particular to Larry Roberts, who read the entire manuscript. I salute all of these scientists as any parasite must salute its host. My thanks go to:

Greta Smith Aeby
Jonathan Baskin
Nancy Beckage
George Benz
Manuel Berdoy
Jeff Boettner
Daniel Brooks
Janine Caira
Dickson Despommiers
Andrew Dobson
Thomas Eickbush
Gerald Esch
Donald Feener

Michael Foley
Scott Gardner
Matthew Gilligan
Bryan Grenfell
Iah Harrison
Hans Herren
Eric Hoberg
Jens Høeg
Peter Hotez
Stephen Howard
Frank Howarth
Michael Huffman
Hillary Hurd

Todd Huspeni
Mark Huxham
John Janovy
Daniel Janzen
Aase Jesperson
Pieter Johnson
Martin Kavaliers
Christopher King
Jacob Koella
Stuart Krasnoff
Armand Kuris
Kevin Lafferty
Curtis Lively

Acknowledgments

Philip LoVerde
David Marcogliese
Scott Miller
Katherine Milton
Anders Møller
Janice Moore
Thomas Nutman
Jack O'Brien
Richard O'Grady
Norman Pace
Edward Pearce
Barbara Peckarsky

Kirk Phares
Stuart Pimm
Ramona Polvere
Mickey Richer
Larry Roberts
David Roos
Mark Siddall
Joseph Schall
Phillip Scott
Andreas Schmidt-
    Rhaesa
Biola Senok

Michael Strand
Michael Sukhdeo
Suzanne Sukhdeo
Richard Tinsley
John Thompson
Nelson Thompson
Mark Torchin
Joel Weinstock
Clinton White
Marlene Zuk

Also, thanks to David Berreby for some insights on history, Jonathan Weiner for making the worm connection, Grace Farrell for hosting the parasite movie marathon and otherwise tolerating a strange obsession, Eric Simonoff for recognizing fertile gruesomeness when he saw it, and my editor, Stephen Morrow, who, as ever, makes it all happen.

Index

Index

INDEX

Index

INDEX

Index

Merozoites, 12
Microbes, 10–11, 128–29
Milton, Katherine, 179
Mites, xxi, 127, 152–53, 154, 183
Mitochondria, 121, 122, 125, 130
Monogeneans, 52
Moore, Janice, 95–96
Moose, 105, 111
Morris, Kimo, 107–08
Mosquitoes, xix, 12, 13, 38, 50, 64, 89, 90–91, 196, 212
Mother(s)
  and fetus, 135–37
Moths, 152–53, 232
Movies and television shows, 112–15, 116
Multicellular organisms, evolution of, 131, 159–60
Multi-membraned Body, 121
Mustard plant, 83
Mutations, 20, 77, 120, 151, 161, 186, 188
  in hemoglobin, 196–97
  Plasmodium, 208

National Parasite Collection, 137–40, 147, 156
Natural Law in the Spiritual World (Drummond), 17–18
Natural selection, 14, 15, 21, 128, 129, 150, 151, 152, 158, 159, 185, 186, 208
Natural world, relationship to, 116–17, 244
Nematodes, xx, 14, 27–28, 31, 33, 45–46, 84–86, 105, 137, 138, 147–49, 155
  in plant roots, 45–46
  reproduction, 168–69
  and wasps, 153–54
Neurotransmitters, 97–98, 107, 116

Oceans, viruses in, 104
Odor, 25, 99, 111, 180
  anxiety-causing, 93
  in human communication, 199
  illusion of, 134, 135

in mate selection, 175
in plant's defense against parasites, 181
Onchocerca volvulus, xv, 205–06
Origin of Species, The (Darwin), 157–58, 185
Ovalocytosis, 197, 199
Overgrazing, 243

Parasite inventory, 237–40
Parasite navigation, 24, 26–35, 46
Parasites, xvi, xxi
  competition among, 47–50
  contempt for, 15, 21–22, 79, 115–16
  definition, xi, 1–2, 125–26
  domesticating, 212
  eradication of, 205–06, 211–15
  failure to control disease-causing, 203–15
  fear of, 115–16
  getting to new host(s), 51–54
  history of, 122–47, 155–56
  home(s) in body(ies), 35–43
  hosts altered by, 81, 82–83, 91–92, 94–100, 105–10, 149, 245
  humans shaped by, 191–99
  knowledge of, 1–22
  leaving hosts, 50–51, 83–84
  meaning of, 245
  as sign of ecological health, 241–43
  stages/generations of, 6–9
  study of, xviii–xxi, 26–35
Parasitic stability (theory), 243–44
Parasitic Wealth or Money Reform (Brown), 19
Parasitism, 17–19, 125, 126, 128
  as genetic interests, 136–37
  in multicellular organisms, 159
  social, 133–35
Parasitologists, xiv–xv, xxi, 19, 24, 27, 66–67, 103, 213
  on manipulators, 97, 98
  studying blood flukes, 72, 73, 75
  studying manipulations, 94, 96
  studying nematodes, 148
  studying virulence, 154
  tools of, 119–20

INDEX

Index

INDEX